Science and Fiction

Science and Fiction – A Springer Series

This collection of entertaining and thought-provoking books will appeal equally to science buffs, scientists and science-fiction fans. It was born out of the recognition that scientific discovery and the creation of plausible fictional scenarios are often two sides of the same coin. Each relies on an understanding of the way the world works, coupled with the imaginative ability to invent new or alternative explanations—and even other worlds. Authored by practicing scientists as well as writers of hard science fiction, these books explore and exploit the borderlands between accepted science and its fictional counterpart. Uncovering mutual influences, promoting fruitful interaction, narrating and analyzing fictional scenarios, together they serve as a reaction vessel for inspired new ideas in science, technology, and beyond.

Whether fiction, fact, or forever undecidable: the Springer Series "Science and Fiction" intends to go where no one has gone before!

Its largely non-technical books take several different approaches. Journey with their authors as they

- Indulge in science speculation – describing intriguing, plausible yet unproven ideas;
- Exploit science fiction for educational purposes and as a means of promoting critical thinking;
- Explore the interplay of science and science fiction – throughout the history of the genre and looking ahead;
- Delve into related topics including, but not limited to: science as a creative process, the limits of science, interplay of literature and knowledge;
- Tell fictional short stories built around well-defined scientific ideas, with a supplement summarizing the science underlying the plot.

Readers can look forward to a broad range of topics, as intriguing as they are important. Here just a few by way of illustration:

- Time travel, superluminal travel, wormholes, teleportation
- Extraterrestrial intelligence and alien civilizations
- Artificial intelligence, planetary brains, the universe as a computer, simulated worlds
- Non-anthropocentric viewpoints
- Synthetic biology, genetic engineering, developing nanotechnologies
- Eco/infrastructure/meteorite-impact disaster scenarios
- Future scenarios, transhumanism, posthumanism, intelligence explosion
- Virtual worlds, cyberspace dramas
- Consciousness and mind manipulation

More information about this series at http://www.springer.com/series/11657

Andrew May

The Science of Sci-Fi Music

 Springer

Andrew May
Crewkerne, UK

ISSN 2197-1188   ISSN 2197-1196   (electronic)
Science and Fiction
ISBN 978-3-030-47832-2   ISBN 978-3-030-47833-9   (eBook)
https://doi.org/10.1007/978-3-030-47833-9

This Springer imprint is published by the registered company Springer Nature Switzerland AG.
The registered company address is: Gewerbestrasse 11, 6330 Cham, Switzerland

Contents

Alien Sounds

During the 1950s and 60s, movie soundtracks created a whole new language to represent alien, futuristic and supernatural themes. Emphasizing dissonance, rhythmic ambiguity and unfamiliar electronic sounds, this language had surprising similarities to the avant-garde classical music of the same period. In this chapter we take a brief look at the origins of these new sounds in the music of the early 20th century, then go on to discuss their use in classic movies such as *The Day the Earth Stood Still*, *Forbidden Planet*, *Curse of the Werewolf* and *Planet of the Apes*—not forgetting Stanley Kubrick's inspired use of pre-existing avant-garde compositions in *2001: A Space Odyssey* and *The Shining*.

Air of Other Planets

What does "alien" or "supernatural" music sound like? Everyone knows the answer, thanks to countless science fiction and horror movies. It's basically a negation of the things that make conventional music—of all cultures—sound "natural" or "down-to-earth": repetitive rhythms, simple harmonies, easily discernible tunes. Replace those with irregular or non-existent rhythms, dissonant chords and a seemingly random sequence of notes—perhaps played on an unfamiliar, electronic instrument—and what you're left with, almost by definition, sounds alien.

It's a simple formula, and it works. As popular culture historian Mathew Bartkowiak says in a book on sci-fi film music, *Sounds of the Future* (2010):

© The Editor(s) (if applicable) and The Author(s),
under exclusive licence to Springer Nature Switzerland AG 2020
A. May, *The Science of Sci-Fi Music*, Science and Fiction,
https://doi.org/10.1007/978-3-030-47833-9_1

"one wave of the hand in front of the theremin or one chord played on the piano can create a shorthand that speaks to us more than any screenwriter could hope to affect with dialogue" [1].

It works so well, in fact, that it's become a Hollywood cliché. In another chapter in same book, Lisa Schmidt writes that:

> The conventions of representing alien beings, times and/or spaces through the electronic, the atonal or dissonant have been more or less continuous, from roughly 1950, when American sci-fi begins as a film genre, to the present [2].

The word "atonal" in that quote sounds like a synonym of "tuneless", but actually it has a more specific meaning. To quote from a BBC website aimed at high-school music students, "atonal music is not related to a tonic note and therefore has no sense of key" [3].

If you happen to be a high-school music student—or otherwise have a musical background—you'll understand that sentence, but to anyone else it looks pretty mystifying. Fortunately we won't be needing much music theory in this book, because the kind of music we're interested in tends to break the rules rather than follow them. Still, it's worth a brief detour at this point to explain some of the basics of those rules.

To start with, almost all western music is based around 88 distinct pitches corresponding to the black and white keys of a standard piano. As we'll see in the next chapter ("Musical Mathematics"), these pitches correspond to different audio frequencies, which can be measured in Hertz, while musicians refer to them by time-honoured letter-codes. But there's a third way to describe musical pitches, which is easier to get to grips with than either frequencies or letter-codes. It's the way a MIDI keyboard encodes pitches.

If you think about it, a key on a MIDI keyboard—as opposed to an acoustic instrument—doesn't actually produce a sound, it just sends a number to a computer program. It's a very simple number—an integer—which goes up in unit steps from one key to the next. The step size, in musical jargon, is called a "semitone" (one of the two music-theory terms we're really going to need in this book).

Somewhere near the middle of the keyboard there's a white note that musicians call "Middle C". Pressing it generates the number 60—which is all you need to know in order to work out the integers for all the other notes. The black note immediately to the right of Middle C is 61, then the white note after that is 62—and so on for another ten notes up to 72.

At this point, you'll notice that the pattern of keys starts to repeat—the ones around 72 look just like those around 60. That's not an accident, because the note generated by key 72 sounds very similar to Middle C, but in a higher register. In musical jargon, it's an "octave" higher (that's the other music-theory term we can't avoid using).

While integers like 60 and 72 are useful in telling us exactly how high or low a note is, as far as musical harmony is concerned there's no difference between the two. That suggests that, in mathematical terms, we should be using "modulo-12" arithmetic—just considering the remainder when we divide an integer by 12. In this approach, 60 and 72 both become zero, while 61 and 73 become 1, and so on. This reduces the whole of western music to just 12 "pitch classes" (PCs), as shown by the red numbers in Fig. 1.

This formalism is the basis for musical set theory, which will be discussed in more detail in the next chapter. The important thing for now is to understand that "tonal music"—the great bulk of western music, both classical and popular—is based around specific subsets of pitch classes called "scales" and "chords".

As many readers will already know, the C major scale only uses the white keys of the piano—the seven PCs (0, 2, 4, 5, 7, 9, 11). The sound of a scale isn't determined by the PCs themselves, so much as by the intervals between them. If you add a constant number of semitones—modulo-12, of course—to each PC, you get another major scale. For example adding 3 semitones gives you the scale musicians call E-flat: (3, 5, 7, 8, 10, 0, 2).

There are other rules—either written or unwritten—that specify how the notes of a scale should be combined "horizontally" (i.e. played consecutively) or "vertically" (played simultaneously). The latter is where the concept of chords comes in. Most people are familiar with these because they're what rhythm guitarists play in rock music, but in more subtle ways chords provide the basis for vertical harmonies in all kinds of music. To oversimplify a huge

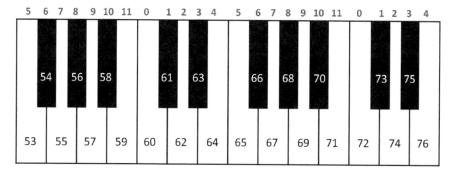

Fig. 1 The central part of a MIDI keyboard, showing the integer returned by pressing each key, and its "mod-12" equivalent in red

subject, a chord is a subset of 3 or 4 PCs, most elements of which are 3 or 4 semitones from their nearest neighbours—such as the C major chord (0, 4, 7) or the G7 chord (7, 11, 2, 5).

Within the rigorous tonal framework that prevailed prior to the 20th century, composers had limited options when it came to depicting the supernatural. One much-used trope was the so-called "tritone"—an interval of three whole tones, or six semitones. We've just seen one, because the G7 chord contains PCs 5 and 11 (notes F and B in traditional musical parlance), which are six semitones apart. The presence of other notes in G7 gives it quite an earthy sound, but playing the two tritone notes on their own produces a much spookier effect. It's much used, for example, in the segment of Hector Berlioz's *Symphonie Fantastique* (1830) called "Dream of the Witches' Sabbath". The horror website *Emadion* describes the tritone as "the undisputed protagonist" of that movement [4].

Since there are 12 semitones—six whole tones—in an octave, a tritone is exactly half an octave. That sounds innocent enough, but it's an interval many musicians consider the most "dissonant" of all. There's a good scientific reason for this, which we'll see in the next chapter when we look at the relationship between musical pitches and frequencies of vibration. For the moment, suffice it to say the tritone has sinister connotations going back long before Berlioz.

In mediaeval times, the tritone was known as *diabolus in musica*, or "the devil in music". More recently—particularly in the context of heavy metal music, where it's used a lot—it's been dubbed "the devil's chord". Here is musicologist John Deathridge on the subject:

> The Devil's interval enjoyed great popularity among composers in the 19th century, when you have got lots of presentations of evil built around the tritone. It can sound very spooky. It depends on how you orchestrate… Wagner's *Götterdämmerung* has one of the most exciting scenes—a pagan, evil scene, the drums and the timpani. It's absolutely terrifying, it's like a black mass [5].

Another composer who put the tritone's sinister sound to good use was Franz Liszt, a contemporary of Berlioz and Wagner who was even more adventurous musically. His *Dante Sonata* of 1849 is essentially a musical picture of Hell, as described in Dante's classic poem *Inferno*. As Liszt's biographer Derek Watson says, "a broad but vivid impression of *Inferno* is achieved through breathtaking use of keyboard effect and structural use of the tritone" [6].

Several years later, Liszt took on another classic supernatural subject in his *Faust Symphony* (1857), based on the legend of the scholar who sold his soul to the demon Mephistopheles. In this case, Liszt adopted a different approach. Instead of using the well-established "spooky" interval of the tritone, he broke with tonal tradition in a way that had no precedent at all. The opening bars of the *Faust Symphony* are completely atonal, featuring a non-repeating sequence of all 12 PCs, in the order 7—11—3—6—10—2—5—9—1—4—8—0. The musical effect is even weirder than the tritone; according to Derek Watson, "commentators have associated it with the mystical and magical aspects of Faust's character" [7].

Watson also quotes a letter that Liszt wrote in 1860, in which the composer describes a kind of super-chord containing all 12 PCs "which will thus form the unique basis of the method of harmony—all the other chords, in use or not, being … the arbitrary curtailment of such and such an interval" [8].

Liszt was probably being facetious—but even so, his suggestion is uncannily prescient of the "set theory" approach to composition that emerged in the 20th century, as we'll see in the next chapter.

Liszt was one of the most remarkable musicians of the 19th century—not only because he was such an innovative composer, but also as arguably the greatest keyboard player the world has ever seen. One of his most famous showpieces, the *Mephisto Waltz*, also came from the Faust legend. Of this "depiction of devilish machinations", Paula Kennedy says, in the notes to one recording:

> For many people this piece provided fresh evidence that the composer himself had dabbled in the supernatural. It was thought that such fiendishly difficult music could be played only by one who, like Faust, had entered into a pact with the Devil, and God-fearing audiences recoiled at its wanton excesses [9].

An updated twist on this idea forms the basis of the 1971 horror movie *The Mephisto Waltz*. Its plot is summarized on the cover of the DVD as follows:

> Alan Alda plays a classical piano player who is on the rise to fame. He befriends a famous pianist who is at death's door. Unknown to Alda, the guy is a Satanist, who arranges to have their souls switch places at his death, so that he can be young again and continue to play piano [10].

The movie features a classic horror soundtrack by Jerry Goldsmith—with not a little help from Liszt. To quote *Gramophone* magazine: "Liszt's piano work of the same name … provides the bedrock upon which Goldsmith builds his nightmarish score, a bleak, uncompromising work that affords no comfort for the unwary listener" [11].

Liszt, however, was just the beginning. Classical music remained essentially "tonal"—in the technical sense defined earlier—throughout the 19th century, and it was only in the first decade of the 20th that atonality started to make its mark in a big way. This is where the roots of many Hollywood horror and sci-fi film scores lie—the new sounds of the early 1900s providing fertile inspiration for composers like Goldsmith at the other end of the century.

Key figures in the shift towards atonality (see Fig. 2) were Arnold Schoenberg (1874–1951) and Anton Webern (1883–1945) in Europe, and Charles Ives (1874–1954) and Edgard Varèse (1883–1965) in America—the latter having emigrated from France.

These four composers had a defining role in 20th century music—and not just in the obvious ways. They're all credited, for example—along with many other people—on the seminal 1966 album *Freak Out* by Frank Zappa and the Mothers of Invention as having "contributed materially in many ways to make our music what it is" [12].

Varèse in particular was a huge influence on Zappa throughout his career. According to his biographer John Corcelli, the first record album he ever bought, at the age of 12, was by Varèse: "He absorbed the liner notes and played the record over and over, not only for himself but also for his friends from high school, who probably didn't appreciate it the same way he did" [13].

Perhaps the greatest single significance of Varèse's music—as with that of Schoenberg, Webern and Ives—is that it enormously expanded the range of what was possible or acceptable in music—or indeed, even what was thought of as music. Varèse himself defined it as "organized sound"—a definition within which almost anything is possible. A theme that will be developed later in this book is the way 20th century music increasingly drew on the world of science, both for inspiration and for practical new techniques. It was a trend

Fig. 2 The four great pioneers of atonal music: Charles Ives, Arnold Schoenberg, Anton Webern and Edgard Varèse (images from Wikimedia Commons)

that can be traced back to Varèse, who said: "music, which should pulsate with life, needs new means of expression, and science alone can infuse it with youthful vigour" [14].

Charles Ives, like Varèse, created a unique sound-world that was decades ahead of its time. In the context of this book, his most significant work was the *Universe Symphony*—composed during the 1920s but never actually performed in his lifetime. It uses a huge orchestra divided into three sections—low-pitched instruments depicting the Earth, high-pitched ones for the Heavens, and percussion providing the "cosmic pulse"—with each sub-orchestra having its own conductor and its own distinct rhythmic pattern.

That may sound like a recipe for chaos, but it's carefully designed chaos, as Johnny Reinhard—who presided over one of the work's first performances in the 1990s—explains: "Ives worked out very clearly a set of cross metres, each with its own polyrhythms, to model programmatically the growth and the intricacy of the universe" [15]. The result, to contemporary ears, is archetypal space music.

Both Varèse and Ives suffered from abysmally low profiles throughout their careers, their various innovations being overlooked or ignored by the musical establishment. That was never a problem for their European counterparts, Schoenberg and Webern, who were among the most influential (if far from "popular", in any sense of the word) composers of the first half of the 20th century.

More will be said about Schoenberg in the next chapter ("Musical Mathematics"), but in the present context one of his most significant early works is his String Quartet No. 2 of 1908. Unusually for what is basically an instrumental piece, the last movement has a part for a soprano singer: a setting of a poem by the German symbolist poet Stefan George called *Entrückung* ("Rapture"). The significance for a book about sci-fi music lies in both the alien sound of the music—described by Schoenberg's biographer Allen Shawn as "some of the eeriest music yet composed by anyone" [16]—and the text itself, which begins "*Ich fühle Luft von anderem Planeten*" ("I feel the air of another planet").

The German word *Luft* simply means "air", in the sense of the stuff you breathe. In the English translation, however, it becomes a kind of pun, because "air" can also mean tune—as in *Londonderry Air*. Strangely enough, the same word also features in a punning title by Ives: *Varied Air and Variations*, which sounds like "very daring variations". A description of that particular piece in the notes to a CD recording give a feel for the uncompromisingly modern approach Ives shared with Schoenberg:

Its core is a series of five diverse variations based on an atonal air, a non-repetitive, awkwardly phrased, hyper-chromatic melody almost impossible to hum, sing, whistle or even remember. All but one of the variations are full of in-your-face dissonance [17].

Going back to the Schoenberg piece—translated into English, the idea of "Airs from Another Planet" is irresistibly evocative from a science-fictional point of view. So irresistible, in fact, that it was used as the title of a short story by Sarah Ash—a trained musician as well as a writer—which appeared in the magazine *Interzone* in May 1994. It revolves around a composer who believes he hears strange music telepathically from an unknown source. "Is it," as a doctor muses in the story, "merely the chaotic rambling of a fractured mind—or is it, as he insists, an emanation from a world far beyond our own?" [18].

Exactly the same phrase, *Airs from Another Planet*, was used as the title of a 1986 CD by the composer Judith Weir [19], who was appointed Master of the Queen's Music—the highest-ranking position for a classical musician in the UK—in 2014.

Long before that, the revolutionary aspects of Schoenberg's String Quartet No. 2 were picked up by his most avid disciple, Anton Webern. In the words of musicologist Paul Griffiths:

> Like Schoenberg, Webern began to explore the air of another planet with the support of words by Stefan George, which he used in the chorus *Entflieht auf leichten Kähnen* ("In swift light vessels gliding") of 1908 [20].

That particular piece by Webern happens to be included on a 1970 album called *2001 Volume Two*, issued by Polydor records in response to the enormous popularity of the soundtrack of *2001: A Space Odyssey*—which, as we'll see later in this chapter, used classical music to great effect. The "Volume Two" disc contains pieces not featured in the movie but written in a similar style—and Webern's *Entflieht auf leichten Kähnen* is among them.

There's another link, albeit a rather tenuous one, between Schoenberg's music and outer space. One of the best known of his early atonal works is *Five Pieces for Orchestra*, from 1909. It had its first London performance in 1912, and one of the people in the audience on that occasion happened to be the composer Gustav Holst—best known, of course, for his orchestral work *The Planets*. Quoting David Lambourn from a 1987 article in *The Musical Times*:

> Holst was particularly impressed with the work and was able to obtain an orchestral score... It is possible to hear echoes of it in his suite *The Planets*, written between 1914 and 1916, the original title of which was, interestingly, *Seven Pieces for Large Orchestra* [21].

Highbrow Music in Lowbrow Movies

While the "atonal" style inaugurated by Schoenberg and Ives never made it into the musical mainstream—either classical or pop—it dominated the avant-garde fringe throughout the 20th century. After the 1940s it was joined by, and inextricably linked to, electronic music, of which Edgard Varèse became one of the first exponents. Also emerging around this time was a highly mathematical alternative to tonality called serialism.[1]

All these topics will be revisited in more detail in the next three chapters. For now, the key point to emphasize is that this kind of avant-garde music was considered highly specialized and "difficult", with a limited audience made up of a subset of professional musicians and intellectuals. It's all the more surprising, therefore, to find it infiltrating its way into the soundtracks of mass-market sci-fi and horror movies of the 1950s and 60s.

It's a point made by genre specialist Lisa Schmidt in an article entitled "A Popular Avant-Garde: The Paradoxical Tradition of Electronic and Atonal Sounds in Sci-Fi Music Scoring":

> To be avant-garde is to be new, modern, pushy—and, by necessity, anything but popular. A genre film is popular by design, virtually incapable of being avant-garde. Certainly science fiction films, consistently popular from the 1950s through to the present, would be difficult candidates for the avant-garde despite the modernist origins of some of their literary and cultural antecedents [2].

In spite of this, as Schmidt say, sci-fi movies have developed a generic tradition of "expressing futurist/alien themes through use of dissonance and/or electronic sounds". It is, as her subtitle points out, a paradox. As she says later in the article, "the same person who has never heard a Charles Ives symphony will consume with relish Howard Shore's deeply atonal scores for David Cronenberg's films" [2].

One of the atonal techniques pioneered by Ives was the use of polychords: two ordinary—but unrelated—tonal chords played simultaneously. For

---

[1] The term serialism dates from the 1940s, and strictly refers to a mathematical generalization of an earlier compositional technique pioneered by Schoenberg. The latter should strictly be called the "12-tone method", but it's often loosely referred to as serialism as well. Because several of the writers quoted follow this looser convention, "serialism" will be used this way in this chapter.

example, the chords D minor (2, 5, 9) and E-flat major (3, 7, 10), each on its own, are perfectly commonplace in tonal music.[2] But putting them together, Ives style, to make (2, 3, 5, 7, 9, 10)—with two semitonal clashes 2-3 and 9-10, and a tritone 3-9—results in a dissonance most composers wouldn't go anywhere near.

Yet that particular chord is familiar to all sci-fi movie fans—as music professor Stephen Husarik explains, in the context of Bernard Herrmann's score for *The Day the Earth Stood Still* (1951), one of the first films to depict the arrival of an alien "flying saucer" on Earth:

> The first chord heard in *The Day the Earth Stood Still* is an unresolved polychord consisting of a D minor triad stacked upon an E-flat triad. With this stridently unresolved polychord we are warned at the outset of the film that unresolved dissonances and harmonic nullification may permeate the entire film. The audience soon realizes that these polychords are not merely a colouristic effect, but serve a structural purpose in the drama [22].

There's a direct link between the appearance of polychords in the movie score and their pioneering use by Charles Ives several decades earlier. As Husarik says, "Herrmann championed the works of Ives in his writings" as well as conducting Ives's music in public concerts.

Another innovative feature of Herrmann's music for *The Day the Earth Stood Still* is its use of electronic instrumentation, in the form of the theremin, invented in the 1920s by Russian physicist—and musician—Léon Thérémin. The instrument uses a well-established process from radio engineering called heterodyning, by which two high-frequency oscillators, tuned to slightly different frequencies, produce a third frequency—in the audible range—at the difference between the first two frequencies.

The same principle is used in another instrument of similar vintage, the Ondes Martenot, developed by the French inventor Maurice Martenot. The latter is a fairly conventional-looking keyboard-style instrument, featured in a number of classical compositions such as Olivier Messiaen's *Turangalîla-Symphonie*—where, according to Paul Griffiths, "Messiaen used its capacity to suggest a voice unearthly in its range, power and wordlessness" [23].

The theremin, on the other hand, not only sounds sci-fi but looks it (see Fig. 3). It's played without any physical contact by the performer, via two

---

[2] Readers with a musical background will know immediately what terms like "D minor" and "E-flat" mean. For those who don't, integer notation is probably more helpful. Just remember we're using mod-12 arithmetic, and intervals 3, 4 and 5 are considered harmonious, while 1, 2 and 6 are "dissonant".

Fig. 3 Alexandra Stepanoff playing the theremin on NBC radio in 1930 (public domain image)

antennas that sense the proximity of the thereminist's hands—one controlling frequency and the other volume.

The eerie sound of the theremin was something of a Hollywood cliché in the 1950, as Lisa Schmidt explains:

> The sonority of the theremin became quickly associated with themes of mental instability, threat, and finally the alien or monstrous. The first science fiction film to feature the theremin was *Rocketship X-M* (1950), with a score by Ferde Grofé, but it was incorporation of the theremin in a trio of science fiction classics—*The Day the Earth Stood Still* (1951), *The Thing from Another World* (1951) and *It Came from Outer Space* (1953)… that forever associated it with the alien. Indeed, for a time, the theremin became a staple of science fiction film and television. It is clear that the sound of the theremin imprinted itself on the cultural memories of both artists and film goers, although with access to the next generation of electronic instruments, composers for sci-fi films often resorted to the creation of theremin-like sounds and gestures, rather than the theremin itself [2].

The best-known example of this kind of "pseudo-theremin" sound is the opening theme of the original 1960s *Star Trek* series, where—ironically for such a futuristic show—the electronic instrument is impersonated by a female vocalist.

In the early days of electronic music, instruments that could be played in real-time, like the theremin, were in the minority. Most electronic music had to be produced slowly and laboriously "offline", using signal generators and magnetic tape. It's a subject we'll deal with in detail in a later chapter ("The Electronic Revolution"), covering the experimental work of Edgard Varèse and his younger followers such as Karlheinz Stockhausen and Milton Babbitt. Just as pioneering, however, was the soundtrack of *Forbidden Planet* (Fig. 4)—a 1956 film described by Ken Hollings in a BBC radio documentary as "a low budget Freudian retelling of Shakespeare's *Tempest* set in outer space" [24].

The film's score—the first to consist of nothing but electronically produced sounds—was created by husband-and-wife team Louis and Bebe Barron. They came very much from the avant-garde scene, living and working in New York's Greenwich Village. There, in the early 1950s, they met John Cage,

Fig. 4 *Forbidden Planet* (1956) featured an all-electronic soundtrack by avant-garde composers Louis and Bebe Barron (public domain image)

a pioneer of new music who also subscribed to the philosophy of Zen Buddhism. This advocated spontaneity and "non-intentional" creativity—ideals that Cage strove to achieve in his music. As Hollings says in the BBC documentary:

> Louis and Bebe were recruited by Cage to create an early form of synthetic music. Their task was to create a library of individual recordings chopped into tiny slices of tape by Cage and sorted and arranged according to chance operations [24].

The resulting composition became known as *Williams Mix*, after the African-American architect Paul Williams who funded the undertaking. As musicologist Paul Griffiths explains:

> *Williams Mix*, of which only a four-minute fragment was ever completed, required a host of coin tosses to determine the kinds and lengths of sounds to be spliced together onto eight simultaneous tracks, each sound belonging to one of six categories—"city sounds", "country sounds", "electronic sounds", "manually produced sounds including the literature of music", "wind-produced sounds including songs" and "small sounds requiring amplification to be heard with the others"—and subjected or not to control of frequency, overtone structure and amplitude. For Cage, the work took him still closer to non-intentionality, since the choices of sounds and controls could be made by other people, following the chance-ordered plan [25].

The Barrons went on to compose soundtracks for several short avant-garde films before tackling *Forbidden Planet*. For this, they used electronic circuits taken from a book on cybernetics by Norbert Wiener, adapted to produce a wide variety of unearthly sounds. But that was just the start of it. As Lisa Schmidt explains, these raw sounds were then "recorded on magnetic tape that was spliced, rerecorded with changes of tape speed and direction, with added echo and reverb" [2]. The result is a strange hybrid of music and non-musical sound effects.

The "forbidden planet" of the title is Altair 4, formerly home to a long-dead race known as the Krell, who left behind relics of their high-tech civilization—including recordings of their music, one of which is played in the course of film. "That recording was made by Krell musicians half a million years ago", the character Dr Morbius explains. In reality, of course, it's another example of Bebe and Louis Barron's work. The BBC's Ken Hollings describes it as "an electrifying and poetic moment … when the phantom sound of Krell technology, Wiener's theory of cybernetics and the Barrons' radical innovations all meet" [24].

Forbidden Planet's score was famously described in the closing credits as "electronic tonalities" rather than "music". That's a great description of its revolutionary new sound, but that's not the reason why it was credited in those terms. Instead, it was a ploy to avoid problems with Hollywood musician's union, of which the Barrons weren't members.

Electronic music was one of two great innovations of the mid-20th century avant-garde movement. The other was serialism—a novel approach to composition which will be explained more fully in the chapters on "Musical Mathematics" and "Scientific Music". Put simply, serialism is a different set of rules for putting notes together—either horizontally to form melodies or vertically to form chords—to those of conventional tonality.

In the world of classical music, experiments in serialism were limited to a small circle of "highbrow" composers. At the same time, however, serial music reached a much larger audience through its use in popular movies like *Curse of the Werewolf* (1961), from Britain's Hammer studios, and the Hollywood blockbuster *Planet of the Apes* (1968), scored by Jerry Goldsmith.

This wasn't a random coincidence but a deliberate aesthetic choice—in the case of Hammer films, at least, who "encouraged an interaction between popular culture and composers who were also active in the world of avant-garde concert music", as David Huckvale explains in his 2008 book *Hammer Film Scores and the Musical Avant-Garde*: "Hammer commissioned a group of leading British composers to write music for its films at a time when the European avant-garde was changing the language and style of music as a whole" [26].

Huckvale goes on to contend that Hammer studios, together with its lower budget rival Amicus, "were extremely influential, far more so than any more overtly intellectual platform, in disseminating a new musical aesthetic".

A central part of this new aesthetic was serialism, which in its broadest sense encompasses everything from Schoenberg's later neo-classical compositions—and those of his close followers such as Alban Berg—in the 1930s, right up to the highly experimental work of younger composers like Stockhausen and Pierre Boulez in the 1950s. Unlikely as it sounds, Hammer films fit in there too, with soundtracks such as the one written for *Curse of the Werewolf* by Benjamin Frankel—a prolific but little-remembered composer who produced eight symphonies as well as scoring dozens of films. Frankel's serialism took a relatively mild form, as Huckvale explains:

> Benjamin Frankel created the first British serial film score in his music for *The Curse of the Werewolf*, starring Oliver Reed in the hirsute and bloody title role;

but Frankel's approach to serialism was very different from the intellectual approach of Pierre Boulez.... Frankel's *Curse of the Werewolf* score resembled even more the general approach of Alban Berg's Violin Concerto, the note row of which also contains elements of tonality [26].

The same view is echoed in *Gramophone* magazine's review of the *Curse of the Werewolf* soundtrack: "we find Benjamin Frankel marrying 12-tone serialism, without abandoning tonality, to the often brief and fragmentary nature of film music with great panache" [27].

The biggest difference between Frankel's score and the great bulk of serial music lies, of course, in the breadth of its audience. As Huckvale says, "more have heard Benjamin Frankel's music for *The Curse of the Werewolf* than have bought a CD of Frankel's own concert works, let alone attended a performance of Berg's Violin Concerto" [26].

While Benjamin Frankel was a relatively minor figure in British classical music, that's not the case with another horror-movie composer, Elisabeth Lutyens (1906–83)—who, according to musicologist Hugh Wood, "has a special importance as the pioneer of 12-note technique in this country" [28]. And quoting from another *Gramophone* review:

> Her reputation is that of a lone, brave modernist voice in the conservative Britain of the 1940s and 1950s … Serialism for her was no dogma or easy route to modernity but a refining process, and with it she distilled a very individual voice [29].

Despite this lofty reputation, Lutyens was a prolific writer for horror movies—chiefly for Hammer's rival Amicus, for which she scored such films as *Dr Terror's House of Horrors* (1965), *The Skull* (1965), *The Psychopath* (1966) and *The Terrornauts* (1967). While these may have been heard many more times than, say, her "Six Tempi for Ten Instruments" (1957), "Quincunx for baritone, soprano and orchestra" (1960) or "Plenum I for piano" (1972), they have a much lower profile in discussions of her work. Quoting Huckvale again:

> *The New Grove Dictionary of Music and Musicians* omits none of Elisabeth Lutyens's operas and symphonic works, but merely mentions, non-specifically, that she also wrote many film, theatre and radio scores. Of course, social forces are at work here that have nothing to do with music and much more to do with the snobbery of genre classification. To some, no doubt, the title of a film such as *Dr Terror's House of Horrors* [Fig. 5] sits uncomfortably next to that of Lutyens' lyric opera, *Isis and Osiris* [26].

Fig. 5 The garish title logo of *Dr Terror's House of Horrors*, reminiscent of a 1950s horror comic, is in stark contrast to its sophisticated musical score by Elizabeth Lutyens (public domain image)

This snobbery didn't extend to Lutyens herself: "naturally outrageous as she was," Huckvale writes, "with her penchant for green nail polish, she enjoyed being known as a Horror Queen". He also quotes her as saying that "the audience for horror films accepted without a murmur shrill atonal music which they would have rejected with irritation in the concert hall" [26].

With their background in contemporary classical music, it's not too surprising to find composers like Frankel and Lutyens incorporating serial techniques in their film scores. It's much more surprising to find one of biggest names in the world of Hollywood music, Jerry Goldsmith, doing the same thing. He's been mentioned once in this chapter already, for his scoring of *The Mephisto Waltz* (1971)—which displays, according to reviewer Jason Comerford, "a lot of Goldsmith's revolutionary avant-garde technique" [30].

Goldsmith's score for an earlier and much better-known movie, *Planet of the Apes* (1968), went even further, however, and actually employed serial techniques. The result was impressive enough that it earned Goldsmith an Oscar nomination, with *Gramophone* magazine calling it "a milestone in American film music" [11]. Here's what the Los Angeles Philharmonic website has to say about it:

> Horror, fantasy, and particularly science fiction films have always allowed composers to venture into more outré stylistic modes. *Planet of the Apes* (1968) is one of Goldsmith's key excursions into genre scoring with serial music techniques. "The Hunt", for the scene in which the apes are first seen rounding up

humans, is one of the most ferocious and terrifying action cues ever composed. Cited by Jon Burlingame in his book *Sound and Vision* as "one of only a handful of truly original movie scores", the avant-garde effects for *Planet of the Apes* were all achieved acoustically, i.e. without electronics [31].

These acoustic effects were groundbreaking in their own right, as Goldsmith's Wikipedia entry makes clear:

When scoring *Planet of the Apes*, Goldsmith used such innovative techniques as looping drums into an echoplex, using the orchestra to imitate the grunting sounds of apes, having horns blown without mouthpieces, and instructing the woodwind players to finger their keys without using any air. He also used steel mixing bowls, among other objects, to create unique percussive sounds [32].

The result, in the words of *Gramophone* magazine, is "a work that is both primitive and futuristic" [11]. That's a perfect combination for this particular movie, as cultural anthropologist Cynthia Miller points out:

Goldsmith's use of a vast range of percussive sounds in *Planet of the Apes*, both rhythmic and decidedly arrhythmic, appear designed to harness not only the futuristic imagination, but the evocative power of the most basic, primitive responses—apprehension, panic, flight, anger—as well [33].

Goldsmith went on to use similar techniques in his next sci-fi film score, *The Illustrated Man*, based on a book of that name by author Ray Bradbury. Quoting Miller again:

Goldsmith's score to *The Illustrated Man* (1969) uses some of the same serial techniques found in *Planet of the Apes*, but here he focused the soundtrack on a combination of sterile, emotionless atonal electronics and an earthy, impressionistic theme with fragile, wordless soprano vocals to match the disquieting tone of Ray Bradbury's story collection, from which the film was adapted [33].

Soon after *The Illustrated Man*, Goldsmith collaborated directly with Bradbury on a concert piece called *Christus Apollo*, based on a long poem of that title taken from Bradbury's book *I Sing the Body Electric*. Poems are a rarity in science fiction, and this one is even more unusual for SF in having an overtly pious theme, putting a space-age perspective on the Christ narrative.

Goldsmith's adaptation—a lengthy one, at around 35 minutes—takes the form of a cantata—a traditional way of setting a religious text for solo singers and choir, with orchestral accompaniment. Here's what he says about the work in the CD liner notes:

> In 1969, the California Chamber Symphony asked me to write a cantata based on a text by the celebrated author Ray Bradbury, I was thrilled to be asked since I had a relationship with Ray going back to dramatic radio of the 1950s and later the motion picture *The Illustrated Man*. The cantata was to be a large piece—orchestra, choir, mezzo-soprano, and narration. Although the text is quite spiritual, I elected to compose the piece using the 12-tone system. I feel there is a great relationship between impressionism and dodecaphonicism and that was the musical language I wanted for *Christus Apollo* [34].

The intrinsically "sci-fi" sound of 12-tone music is a perfect match to Bradbury's text, which despite Goldsmith's description of it as spiritual has quite a spacey feel—as you can see from the following brief extract:

> Among the ten trillion beams
>   A billion Bible scrolls are scored
>   In hieroglyphs among God's amplitudes of worlds;
>   In alphabet multitudinous
>   Tongues which are not quite tongues
>   Sigh, sibilate, wonder, cry:
>   As Christ comes manifest from a thunder-crimsoned sky.
>   He walks upon the molecules of seas
>   All boiling stews of beast
>   All maddened broth and brew and rising up of yeast.
>   There Christ by many names is known [34].

As prominent as avant-garde music was in the sci-fi and horror films of the 1950s and 60s, it didn't last. By the time we reach John Williams's score for *Star Wars* (1977)—not to mention its numerous progeny and imitators—the sound had switched from theremins and 12-tone harmonies to something more reminiscent of a Wagner opera, or the late 19th century symphonies of Anton Bruckner or Gustav Mahler.

At the same time, with the advent of practical and versatile commercial synthesizers, electronic music made a sudden jump from the radical fringe to the everyday mainstream. As Lisa Schmidt says in her article on "A Popular Avant-Garde":

> As the end of the 20th century approached, the electronic-atonal convention morphed into something much less obviously alien. Its synthetic entities could be found among musical scores that invoked equally, or even to a greater extent, the classical tradition. For example the Vangelis score for *Blade Runner* (1982)

uses a synthesizer not only to create ambient sounds but also to simulate an entire orchestra [2].

In another chapter of the same book, film critic Matthias Konzett argues that the ambiguity of the *Blade Runner* score may be an intentional reflection of the film's theme—the artificial impersonating the natural:

> It appears that the synthesized soundtrack by Vangelis, electronically mimicking the romantic music of Wagner and Bruckner, likewise points to the constructed nature of all sound, and hence all reality. The presumed anteriority of orchestral music as quasi-natural sound over electronic music as artificial sound is ultimately questioned, just as the difference between replicant and human being becomes increasingly indistinguishable [35].

The whole subject of synthesizer music will be discussed in more detail in a later chapter ("The Electronic Revolution"), including the seminal soundtrack produced by Wendy Carlos for Stanley Kubrick's film *A Clockwork Orange* (1971). At this point, however, it's time to look at another Kubrick classic, *2001: A Space Odyssey* from 1968.

The *2001* soundtrack is particularly noteworthy, not just because it includes a large helping of avant-garde music, but because it wasn't composed specially for the movie at all, but for the concert hall. As such, it's as highbrow as music gets—but *2001* can hardly be called a lowbrow movie, so it doesn't belong in this section "Highbrow Music in Lowbrow Movies". There's so much to say about it, anyway, that it deserves a whole section of its own.

György Ligeti's Space Odyssey

The entire soundtrack of *2001: A Space Odyssey* consists of pre-existing classical music. Scenes depicting human activities in space use uncomplicated dance music—most famously Johann Strauss's *Blue Danube* waltz, but also Aram Khachaturian's music for the ballet *Gayane*. Natural phenomena—specifically the alignment of Sun, Moon and Earth at the start of the film—are represented by the dramatic opening bars of Richard Strauss's tone poem *Also Sprach Zarathustra*—now so inextricably linked to the movie that it's more commonly referred to simply as the "*2001* theme tune".[3]

---

[3] The two Strausses were unrelated. Johann Strauss (1825 – 1899) was a highly popular Austrian composer of light music, while the less familiar Richard Strauss (1864 – 1949) was a German composer of the Wagnerian school. The latter Strauss has another connection with science fiction besides 2001, as we'll see in the final chapter, "Speculations on a Musical Theme".

Of more interest in the present context are the scenes involving the myste-rious alien monolith, because that's where the avant-garde sounds comes in. It's all very strange, other-worldly music—so much so that it's often referred to, erroneously, as "electronic". Even film critic Matthias Konzett, quoted a moment ago, makes this mistake when he refers to "the dissonant overture of electronic music preceding the film's opening scene" [35].

In fact all this music—the work of Hungarian composer György Ligeti (1923—2006)—uses conventional classical forces of orchestra, choir and solo vocalists. The pieces heard in *2001* date from 1961–66, when Ligeti (Fig. 6) had returned to purely acoustic composition after several years of electronic experimentation. Nevertheless, the latter experience affected the way he employed conventional resources, as John R. Pierce, in his book on electronic music, *The Science of Musical Sound*, explains:

> Ligeti abandoned the limited and difficult electronic means of tone production available at that time, but his music shows that he is acutely aware of the subtle qualities of electronic sounds and of the musical value of the sophisticated con-trol of sound quality [36].

There are two aspects to the characteristic Ligeti sound. First there's the wide range of sound textures, technically referred to as "timbres", that he gets out of his instrumentalists and singers. Then there's the way he weaves these

Fig. 6 György Ligeti, whose music features prominently in *2001: A Space Odyssey* (public domain image)

sounds together via "polyphony", to use another technical term. In both cases, Ligeti drew on his electronic experience to create a sound that was uniquely his own. To quote his own words:

> From 1958 on, I developed structures based on "polyphonic nets", music with a web-like character. Significant impetus came from my apprenticeship in the Electronic Music Studio in Cologne from 1957 to 1958. I transferred concepts of sound synthesis to an orchestra [37].

He goes on to describe one specific effect, which can be heard in several places in the music Kubrick chose to include in *2001*:

> I deliberately exploited the effect of combination tones—specifically, difference tones—pitches not actually fingered by the instrumentalists, but which result from them playing together. I heard this acoustic phenomenon as a young child, when several girls with high voices would sing Hungarian folk songs in less than perfect unison. It was an amazing sound, much lower than the one being sung or played, and one does not know from which direction it is coming. As a child I was baffled. It was not until I worked in the Electronic Music Studio that I learned how to create such sounds on purpose. These difference tones are not a purely physical phenomenon, but a psychoacoustic one … some people plug their ears or run from the hall, because it is not entirely pleasant [37].

The work that forms the "overture" to *2001* that Konzett referred to, and reappears during the iconic sequence after astronaut Bowman passes through the Star Gate, is one of Ligeti's best known works, *Atmosphères*—an orchestral composition from 1961 which lasts around nine minutes in total (only a portion of it is used in *2001*). American composer Steve Reich said of it that "*Atmosphères* is an amazing piece, which defines huge clusters as a compositional technique" [38].

A "cluster", in this context, is the simultaneous sounding of several pitch classes separated by only a few semitones. As with the polychords mentioned earlier, it's a technique that was pioneered by Charles Ives in the early years of the 20th century. Clusters, however, have a completely different acoustic effect from any kind of chords, going beyond the simple concepts of harmony and dissonance. It's easiest to explain using an example.

Take a simple chord like [0, 4, 7], with three well-separated notes—it has a stable, harmonious sound. If we then add a fourth note separated by a single semitone from one of the first three notes—for example [0, 4, 6, 7]—the chord becomes dissonant. People often describe dissonances as "unpleasant", but a better adjective is "unstable"—your ear wants them to move to another, more stable harmony. That's the whole basis of tonal music—and a lot of atonal music too.

But what happens if we scrunch all the notes up close to each other— [4, 5, 6, 7], for example? Your ear no longer hears it as a chord but as a single sound, so the concept of "stable" or "unstable" doesn't apply. That's a cluster—and it's a prominent characteristic of Ligeti's music in *2001*.

Another hallmark of Ligeti's music—of that period, anyway—is the almost complete absence of any rhythmic sense. That's something that places it apart from the majority of classical music and the entirety of popular music. Ligeti himself described *Atmosphères* as "static, self-contained music without either development or traditional rhythmic configuration" [38]. As musicologist Paul Griffiths explains:

> Rhythmic movement is eliminated by staggering instrument entries (a technique for which Ligeti introduced the term micropolyphony), emphasizing sustained sounds and avoiding all sense of pulse; harmony is held in suspension by the use of clusters. All these effects of continuity provoke an experience of sound as texture [39].

The result is perfect for the use Kubrick put it to, and it's likely that many moviegoers remain blissfully unaware that the music wasn't written specifically for this context. As Russell Platt wrote in *The New Yorker* magazine: "the Star Gate sequence would have been completely ineffective without Ligeti's dazzling sounds" [40]. Yet Ligeti himself told a friend that he hadn't had "anything cosmic" in mind when composing *Atmosphères* [38].

Kubrick used two other Ligeti pieces in *2001*: the *Requiem* for choir, solo singers and orchestra, and *Lux Aeterna* for unaccompanied choir. Dating from 1965 and 1966 respectively, they were almost brand new—by the standards of classical music—when they appeared in the movie in 1968. Their use in the movie is explained by Robert Cumbow in the liner notes to the soundtrack CD:

> The chord phrases of the *Requiem*, each overriding the one before, create a layering of laments that emerge like the sound of battling winds. This music accompanies the first appearance of the monolith and reflects the furore into which it sweeps the band of man-apes… The less turbulent, more wondrous *Lux Aeterna* accompanies the low-level flight of the Moon bus that carries Dr Floyd to the site of the excavated monolith [Fig. 7], and his team's subsequent approach to the mysterious artifact [42].

Fig. 7 The scenes around the excavated monolith in *2001: A Space Odyssey* were accompanied by Ligeti's *Lux Aeterna* (Matthew J. Cotter, CC-BY-SA-2.0)

The *Requiem* is a setting of the mass used by the Roman Catholic church at funerals—a popular subject among classical composers. Most settings include a section entitled "Lux Aeterna" ("eternal light"), but Ligeti's didn't—an omission he rectified by writing a completely separate piece with that title, as composer Frieder Bernius explains:

> The work was composed for the 16 vocal soloists of the Schola Cantorum Stuttgart. Its conductor Clytus Gottwald had asked the composer for a setting of the *Lux Aeterna* because Ligeti had not set this text, which is the final sentence of the liturgical requiem text, in his *Requiem* which had been completed a short while before [41].

As hinted in the earlier quote by Robert Cumbow, the two pieces are rather different in style. The *Requiem*, aside from the addition of voices, is quite similar to *Atmosphères*, with striking tone clusters and a wide textural range. *Lux Aeterna*, on the other hand, has a gentler, very slow-changing sound—one of the most extreme examples of Ligeti's "micropolyphonic" style. It's a perfect accompaniment to the desolate lunar scenes—"never before or since has music been used to convey such a sense of limitless emptiness", as music journalist Mark Prendergast put it [2].

There's no doubt Stanley Kubrick's decision to use Ligeti's music in *2001* was a stroke of genius, and it's an important part of what makes the film such a unique experience to this day. Nevertheless, there are a couple of unpleasant controversies surrounding the *2001* soundtrack, neither of which reflects particularly well on Kubrick himself.

First, there's the fact that he originally commissioned Hollywood composer Alex North to create an original score for the movie, and then strung him along for many months before ditching him in favour of pre-existing compositions. North told his side of the story in Jerome Agel's book, *The Making of Kubrick's 2001*:

> I was living in the Chelsea Hotel in New York … and got a phone call from Kubrick from London asking me of my availability to come over and score for *2001*. He told me that I was the film composer he most respected, and he looked forward to working together… I flew over to London for two days to discuss music with Kubrick. He was direct and honest with me concerning his desire to retain some of the temporary music tracks which he had been using for the past year. I realized that he liked these tracks, but I couldn't accept the idea of composing part of the score interpolated with other composers. I felt I could compose music that had the ingredients and essence of what Kubrick wanted and give it a consistency and homogeneity and contemporary feel. After having composed and recorded over 40 minutes of music, I waited around for the opportunity to look at the balance of the film, spot the music, etc… Nothing happened. I went to a screening in New York, and there were most of the temporary tracks. Well, what can I say? It was a great, frustrating experience [43].

If Kubrick's treatment of the composer he didn't use was bad, the same was true of the composer he did use. Putting his music in *2001* may have brought Ligeti to the attention of a huge worldwide audience, but it was done without the composer's knowledge. In a recent book about the making of *2001*, author Michael Benson describes how Kubrick discovered Ligeti's music after his wife heard a broadcast of the *Requiem* on BBC radio:

> Eventually Kubrick managed to hear the piece and was captivated by its power. MGM's production office proceeded to make a deal with the Mechanical Rights Protection Society to use it and several other Ligeti works as "background music" for a given price per minute … without the composer's knowledge. When Ligeti went to see the film, he was enraged, wrote music critic Alex Ross. In a letter recently uncovered by the German scholar Julia Heimerdinger, he called the film "a piece of Hollywood shit" [44].

It gets worse. Kubrick not only appropriated Ligeti's music behind his back—and with nothing like the financial reimbursement a Hollywood composer such as Alex North would have received—but he altered part of it too. The situation is alluded to in the liner notes to the soundtrack CD:

> Another use of Ligeti's music occurs in the sequence in the 18th century room in which Bowman lives out his life. The presence of his extraterrestrial zookeepers was suggested by surreal, laughter-like sounds created by altering an excerpt from a Ligeti composition [42].

The composition in question was *Aventures* (1962), a companion piece to *Atmosphères*. It's often reported that Ligeti sued Kubrick over this unauthorized distortion of his music, but according to Benson the matter never reached court—MGM paid up as soon as the matter was brought to their attention.

Even so, Ligeti (who, Kubrick fans will be pleased to hear, eventually came to admire the movie) was never completely happy with the remuneration he received. As he remarked in 1989: "MGM wrote me such nice letters—they said Ligeti should be happy, he is now famous in America". And then in 2001 (the year, not the film), he told an interviewer "I found the way in which my music was used wonderful; it was less wonderful that I was neither asked nor paid" [44].

Fortunately Kubrick didn't make the same mistake when it came to his 1980 horror film, *The Shining*. Quoting *The New Yorker*: "when the director later used Ligeti's music in *The Shining* … he sought permission from Ligeti and paid the composer well" [40]. Ligeti's appearance in that film is a small one, but it coincides with one of the film's most memorable moments. A short excerpt from his orchestral piece *Lontano* is heard when Danny, the young boy at the centre of the story, sees the ghostly twin sisters in the hotel corridor.

Unlike *2001*, only part of *The Shining*'s soundtrack is off-the-shelf classical music. Nevertheless, another avant-garde composer who is featured quite prominently is Ligeti's Polish contemporary Penderecki. As with Ligeti, John R. Pierce noted that "much of Krzysztof Penderecki's music for the conventional orchestra has a distinct 'electronic' sound quality" [36]. It uses a similar combination of tone clusters and complex, arrhythmic textures, but on the whole it's much harsher-sounding, eroding the boundary between music and noise—musicologist Paul Griffiths calls it "texture music" [39].

Penderecki's screechy, spooky music has "horror movie" written all over it, and no fewer than six of his pieces a featured in *The Shining*—albeit in much briefer excerpts than Ligeti's music in *2001*. In many places, viewers could be forgiven for mistaking Penderecki's music for spooky background noises. It's archetypal "organized sound"—to use Edgard Varèse's term—a fact alluded to in the title of one of the most striking Penderecki pieces in the movie, *De*

Fig. 8 A recreation of the "Redrum" door from *The Shining*, from one of the scenes in the movie that was accompanied by Krzysztof Penderecki's music (Carlos Pacheco, CC-BY-2.0)

Natura Sonoris—"on the nature of sound". It appears in another of the film's most iconic scenes, when Danny scrawls "redrum" ("murder" backwards) on one of the hotel room doors (see Fig. 8).

Hopefully the examples in this chapter have illustrated the close affinity between 20th century avant-garde music and the alien, other-worldly or supernatural sensations evoked in sci-fi and horror movies. It's time now to take a more scientific look at why that affinity exists.

References

1. M. J. Bartkowiak (ed.), *Sounds of the Future* (McFarland, North Carolina, 2010), p. 1
2. L.M. Schmidt, *Sounds of the Future* (McFarland, North Carolina, 2010), pp. 22–41
3. BBC Bitesize, *Film and TV Music*, https://www.bbc.co.uk/bitesize/guides/zwdhpv4/revision/2

4. S. Tortiglione, Diabolus in Musica and the Devil's Tritone, *Emadion*, May 2017., https://emadion.it/en/music/diabolus-in-musica-devils-tritone/

5. F. Rohrer, *The Devil's Music*, http://news.bbc.co.uk/1/hi/magazine/4952646.stm

6. D. Watson, *The Master Musicians: Liszt* (Dent, London, 1989), p. 247

7. D. Watson, *The Master Musicians: Liszt* (Dent, London, 1989), p. 275

8. D. Watson, *The Master Musicians: Liszt* (Dent, London, 1989), pp. 184–185

9. P. Kennedy, *Mephisto Waltz* (CD liner notes, Virgin, 1994)

10. *The Mephisto Waltz*, DVD back cover notes (20th Century Fox, 2017)

11. Goldsmith: Planet of the Apes film score, *Gramophone,* https://www.gramophone.co.uk/review/goldsmith-planet-of-the-apes-film-score

12. F. Zappa, *Freak Out* (CD liner notes, Zappa Records, 1987)

13. J. Corcelli, *Frank Zappa FAQ* (Backbeat Books, Milwaukee, 2016), pp. 11–12

14. C. Wen-Chung, *Varèse: The Complete Works* (CD liner notes, Decca, 2004)

15. J. Reinhard, *Charles Ives: Universe Symphony* (CD liner notes, Stereo Society, 2005)

16. A. Shawn, *Arnold Schoenberg's Journey* (Harvard University Press, Massachusetts, 2002), pp. 48–49

17. H. Wiley Hitchcock, *Charles Ives: Piano Sonata No. 2* (CD liner notes, Naxos, 2004)

18. S. Ash, Airs from Another Planet, *Infinity Plus*, http://www.infinityplus.co.uk/stories/airs.htm

19. P. Griffiths, *Modern Music and After* (Oxford University Press, Oxford, 1995), p. 349

20. P. Griffiths, Anton Webern, in *The New Grove Second Viennese School*, (Macmillan, London, 1985), p. 94

21. D. Lambourn, Henry Wood and Schoenberg. *The Musical Times* **128**(1734), 422–427 (August 1987)

22. S. Husarik, *Sounds of the Future* (North Carolina, McFarland, 2010), pp. 164–176

23. P. Griffiths, *Modern Music and After* (Oxford University Press, Oxford, 1995), p. 17

24. BBC Radio 3, *Sound of Cinema: Return of the Monster from the Id*, September 2013. https://www.bbc.co.uk/programmes/b03bfdkj

25. P. Griffiths, *Modern Music and After* (Oxford University Press, Oxford, 1995), pp. 27–28

26. D. Huckvale, *Hammer Film Scores and the Musical Avant-Garde* (North Carolina, McFarland & Co, 2008), pp. 1–5

27. A. Edwards, Frankel: Curse of the Werewolf. *Gramophone.* https://www.gramophone.co.uk/review/frankel-curse-of-the-werewolf

28. H. Wood, English Contemporary Music, in *European Music in the 20th Century*, (Penguin, London, 1961), p. 154

29. M. Oliver, Lutyens Vocal and Chamber Works, *Gramophone*, https://www.gramophone.co.uk/review/lutyens-vocal-and-chamber-works

30. J. Comerford, The Mephisto Waltz by Jerry Goldsmith, *Soundtrack.net*, https://www.soundtrack.net/content/article/?id=26

31. *Planet of the Apes: The Hunt*, Los Angeles Philharmonic, https://www.laphil.com/musicdb/pieces/2890/planet-of-the-apes-the-hunt
32. Wikipedia article on Jerry Goldsmith, https://en.wikipedia.org/wiki/Jerry_Goldsmith
33. C.J. Miller, *Sounds of the future* (McFarland, North Carolina, 2010), pp. 210–222
34. J. Goldsmith, *Christus Apollo* (CD liner notes, Telarc, 2002)
35. M. Konzett, *Sounds of the Future* (McFarland, North Carolina, 2010), pp. 100–115
36. J.R. Pierce, *The Science of Musical Sound* (Scientific American Books, New York, 1983), p. 15
37. G. Ligeti, *Chamber Music* (CD liner notes, Sony 1998)
38. N. Simeone, *Ligeti: Atmosphères* (CD liner notes, Deutsche Grammophon, 2012)
39. P. Griffiths, *Modern Music and After* (Oxford University Press, Oxford, 1995), p. 136
40. R. Platt, Clarke, Kubrick, and Ligeti: a Tale, *The New Yorker*, August 2008., https://www.newyorker.com/culture/goings-on/clarke-kubrick-and-ligeti-a-tale
41. F. Bernius, *György Ligeti: Requiem* (CD liner notes, Carus, 2017)
42. R.C. Cumbow, *2001: A Space Odyssey Soundtrack* (CD liner notes, Sony Music, 1996)
43. J. Agel, *The Making of Kubrick's 2001* (Signet, New York, 1970), pp. 198–199
44. M. Benson, *Space Odyssey: Stanley Kubrick, Arthur C. Clarke and the Making of a Masterpiece* (Simon & Schuster, New York, 2019), pp. 360–362

Musical Mathematics

It's often claimed that music and mathematics are closely connected, although given their different cultural associations many people find this hard to believe. In this chapter, we look at a couple of ways music really can be considered mathematical. First there's the connection, known since ancient times, between musical pitch intervals and the ratios of acoustic frequencies. As interesting as this is to scientists, it's little help to musicians because intervals are perceived as differences, not ratios. An alternative formulation, musical set theory, is more useful in this respect. Among other things, it helps to explain why the "sci-fi" sounds described in the previous chapter sound as alien as they do.

The Harmony of the Spheres

Music has been intimately tied to mathematics—in the mind of mathematicians, if not musicians—at least since the 6th century BCE. That was the time of the Greek philosopher Pythagoras, best remembered for his theorem about right-angled triangles ("the square of the hypotenuse is equal to the sum of the squares of the other two sides"). Pythagoras was also intrigued by the mathematical relationships between musical pitches, as Isaac Asimov—best known for his science fiction, but also a prolific writer on science fact—explains:

> He found that the strings of musical instruments delivered sound of higher pitch as they were made shorter. Furthermore he found that the relationship of

© The Editor(s) (if applicable) and The Author(s),
under exclusive licence to Springer Nature Switzerland AG 2020
A. May, *The Science of Sci-Fi Music*, Science and Fiction,
https://doi.org/10.1007/978-3-030-47833-9_2

pitch could be simply correlated with length. For instance, if one string was twice the length of another, the sound it emitted was just an octave lower. If the ratio of the strings was three to two, the musical interval called a fifth was produced, and if it was four to three, the interval called a fourth was produced [1].

As we saw in the previous chapter, an octave is conventionally divided into 12 semitones, which we can think of as pitch classes (PCs) 0–11. In these terms, the scale of C major, for example, is represented by the 7 PCs (0, 2, 4, 5, 7, 9, 11). The fifth note of the scale is PC 7, so the interval between 0 and 7 is what musicians mean by a "fifth". By the same logic, a fourth is the interval from 0 to 5.

So how do the string-length ratios mentioned by Asimov—1.5 for a fifth and 1.333 for a fourth—relate to intervals of 5 and 7 semitones? It turns out the relationship is, to a good approximation, a logarithmic one:

- The logarithm to base 2 of 1.333 is very close to 5/12
- The logarithm to base 2 of 1.5 is very close to 7/12

This implies that our subjective experience of music is logarithmic. Rather than hearing the ratio of two different string lengths as a ratio, we hear it as the logarithm of that ratio. As meaningful as that sentence is to people with a mathematical background, it just sounds like gobbledegook to most musicians. It's why they're often vehemently opposed to any suggestion that music might be "mathematical", because talking in terms of string ratios and logarithms seems so irrelevant to the practical experience of music.

On the other hand, for Pythagoras (Fig. 1)—and other mathematicians of the ancient and mediaeval world—his discovery had a profound and far-reaching effect, as Eli Maor explains in his book *Music by the Numbers*:

> Pythagoras saw in this a sign that nature itself—indeed, the entire universe—is governed by simple numerical ratios. "Number rules the universe" became the Pythagorean motto, and it would dominate scientific thought for the next 2,000 years [2].

For Pythagoras, the whole cosmos was a kind of gigantic musical instrument. According to physicist and science historian Jean Charon:

> Pythagoras built up his cosmology solely by incorporating numbers in the way they were introduced in the harmony of a string instrument. To him, the order of the universe seemed to prove that the stars and planets travelled along their

Fig. 1 A 15th century woodcut showing Pythagoras experimenting with a musical instrument (public domain image)

orbits not in a haphazard manner, but in such a way that their movements created a celestial harmony just as the lengths of the strings created a musical harmony.

This wasn't just an analogy, in Pythagoras's view, but actual fact—as Charon goes on to explain:

Air was thought to fill all the heavens; in disturbing the air, these various bodies produced musical notes, just as the strings of an instrument will do when plucked. The concert due to these moving bodies was called the harmony of the spheres [3].

Many centuries later, the idea of the music of the spheres was taken a step further by the German astronomer Johannes Kepler, best known for his three laws of planetary motion. The third of these ("the square of the orbital period is proportional to the cube of the semi-major axis") is the high point of his five-volume work *Harmonices Mundi*—"the harmony of the world"—which appeared in 1619. But it doesn't make an appearance until the final volume, "after an outline of various mathematical and musical ideas in the first four volumes", as historian David Love says. Even then, quoting Love again:

> The third law is a far from prominent feature in volume 5 of the book. In most of it, Kepler gives full rein to his mystical and numerological outlook and introduces or revisits ideas that we now know are just plain wrong… He returns to an idea dating back to Pythagoras and looks for musical relationships in the ratios of planetary orbits where none exists. He then spends far more time on these relationships than on his third law. Page after page is filled with tortuous and convoluted arguments trying to explain exactly what the musical relationships are [4].

To Pythagoras, the idea of the harmony of the spheres was an attempt to make sense of the cosmos in proto-scientific terms. By Kepler's time science had moved on, and it no longer had any real use for the Pythagorean model. For him, the music of the spheres fulfilled a religious rather than scientific role, because he viewed the universe as an expression of God's will. As S. D. Tucker writes in his book *Space Oddities*:

> In his strange 1618 text *The Harmony of the World* … dealing initially with the concept of harmony in mathematics and geometry, Kepler goes on to apply similar ideas to the worlds of music, astrology and astronomy too, finding certain key mathematical and geometrical ratios common to each field, and concluding that these were the pure harmonies and proportions which God made use of when constructing his universe. The end effect is to present the entire cosmos as being one big holy mathematical song, with the sense of beauty we feel when listening to harmonic music being due to an unconscious reconnection occurring between the microcosm of one's mind and the macrocosm of God. "The heavenly motions are nothing but a continuous song for several voices," said Kepler, these voices being the planets [5].

Kepler's book even includes examples of the "music of the spheres" in traditional notation (see Fig. 2). As silly as this may look to us today, there's actually some perfectly good science underlying it—closely related, in fact, to Kepler's laws of planetary motion. It would be straying off topic to go into the details here, but it's a subject we'll come back to in a later chapter ("Scientific Music").

If Pythagoras's concept of the music of the spheres hasn't been borne out by modern science, his discovery about the relationship between string lengths and musical pitches is still an important one. In a roundabout way, as we'll see shortly, it even paved the way to electronic synthesizer technology.

Fig. 2 An excerpt from Kepler's book *Harmonices Mundi*, illustrating his conception of the "music of the spheres" (public domain image)

In scientific terms a sound is an acoustic wave travelling through the air, or any other medium. Inside the fluid-filled inner ear, this wave causes thousands of tiny hair-like cells to vibrate, producing the nerve signals that our brains perceive as sound.

What we perceive as the "pitch" of a sound is governed by the frequency of the wave. In the case of a wave produced by a vibrating string, the frequency is determined by the length and tension of the string. Pythagoras correctly observed that, in order to raise the pitch by an octave —in other words to double the frequency—the length of the string has to be halved. However, he also believed that frequency could be doubled by doubling the string's tension—and that's not right. It was only with the rise of experimental science in the 16th century that the correct relationship was discovered, as Pietro Greco explains in his biography of Galileo:

> In order to obtain an octave, the weight you must add to a string in order to vary its tension is not in the 1:2 relationship, as indicated by Pythagoras, but rather in a 1:4 relationship [6].

A mathematician would say that pitch follows a square law with tension, not a linear one. This fact was discovered, not by the famous Galileo Galilei, but by his father Vincenzo—a pioneer of music theory who was almost as scientific in his approach as his son. As Greco puts it, "Vincenzo Galileo was doubtless the first music critic who consulted nature and obtained a general law through an experimental method" [6].

Viewing musical pitches as frequencies helps to explain the fact that, as mentioned in the previous chapter, some pitch intervals sound "harmonious" while others sound "dissonant". Basically, the simpler the numerical ratio between two frequencies, the more harmonious the corresponding interval sounds. Take, for example, the two intervals mentioned in the Asimov quote earlier. They both involve simple ratios of small integers, 3:2 and 4:3, and musicians call the resulting intervals a "perfect fifth" and "perfect fourth" because they're so harmonious sounding.

Now consider the tritone, or "devil's chord", mentioned in the previous chapter. As stated there, this is an interval of exactly half an octave, or six semitones. Knowing that an octave represents a ratio of two, and remembering that the relationship between intervals and frequency ratios is a logarithmic one, mathematicians will see that the tritone must translate to the square root of two. That's a so-called "incommensurable" ratio, which can't be produced simply by dividing one integer by another—and in consequence it sounds much more dissonant to the ear. It's a fact that Galileo—the scientist son, not the musician father—remarked on in his book *Dialogue Concerning Two New Sciences* (1638):

> Especially harsh is the dissonance between notes whose frequencies are incommensurable. Such a case occurs when one has two strings in unison and sounds one of them open, together with a part of the other which bears the same ratio to its whole length as the side of a square bears to the diagonal; this yields a dissonance similar to the tritone or semi-octave [7].

In reality, a vibrating string—or any real-world sound source—produces a whole host of frequencies. If the sound is periodic—i.e. regularly repeating—then the basic frequency of repetition is called the fundamental frequency. This corresponds to the pitch that's most obvious to the ear, and it's the frequency we've been talking about so far. But the only kind of sound that has a fundamental frequency and nothing else is a sine wave, which is typically only ever encountered in electronics laboratories. It isn't a particularly pleasant or musical sound, and paradoxically it's quite difficult for the ear to determine the pitch of a pure sine wave. The reason is that a real musical tone, such as the sound produced by a stringed instrument, contains a whole series of higher frequency components called harmonics.

There's a good explanation of this concept by electronics engineer John R. Pierce, who was a pioneer of computer music—in which context we'll meet him again in the next chapter—as well as an occasional writer of science fiction. He also contributed a series of non-fiction articles, under the pen-name J. J. Coupling, to *Astounding Science Fiction* magazine in the 1940s and 50s. Later in his career Pierce produced a book called *The Science of Musical Sound*, and here's what it has to say about the structure of sound from a scientific point of view:

> The French mathematician Joseph Fourier (1768–1830) invented a type of mathematical analysis by which it can be proved that any periodic wave can be represented as the sum of sine waves having the appropriate amplitude, frequency and phase. Furthermore, the frequencies of the component waves are related in a simple way: they are all whole-number multiples of a single frequency: f_0, $2f_0$, $3f_0$ and so on [8].

Of those frequencies, f_0 is the fundamental, while $2f_0$ is called the 2nd harmonic, $3f_0$ the third harmonic and so on. In general, the harmonics get smaller in amplitude as their frequency increases. That's interesting enough from an analytical point of view, but the real value of Fourier's theory comes from applying it in reverse. Any desired periodic sound can be synthesized by taking a fundamental and adding the necessary harmonics (Fig. 3).

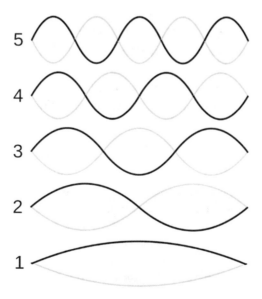

Fig. 3 A periodic musical tone is made up of a series of sine waves, at a fundamental frequency (labelled 1 in this diagram) and various higher harmonics (Wikimedia Commons user Baba66, CC-BY-SA-3.0)

So much for periodic sound waves, but real musical sounds aren't always periodic. A long-held note on a wind instrument may be, but what about a briefly plucked string, or a percussion instrument like a drum or a cymbal? It turns out that such sounds can still be synthesized from sine waves, but not at such neatly defined frequency steps. To quote Pierce again:

> Some sounds used in music consist of frequencies that are not harmonic; that is, are not integral multiples of the lowest frequency... It would sound rather illogical to call these higher frequencies "non-harmonic harmonics", so instead they are called "non-harmonic partials" [8].

The idea that musical sounds can be broken down into their component sine waves is of great interest to scientists—who always like reducing things to their basic elements—and to engineers who want to design electronic instruments. On the other hand, it's of far less interest to musicians themselves, who have traditionally shown very little interest in the scientific or mathematical aspects of their subject. Even before Fourier's time, however, there was a notable exception to this in the form of French composer Jean-Philippe Rameau (1683–1764). As well as writing more than 30 operas, he produced a *Treatise on Harmony Reduced to its Natural Principles*.

Rameau used a scientific understanding of harmonics to rationalize the hitherto completely empirical study of musical chords. As described in the previous chapter, these are sets of notes played simultaneously, such as the C major chord made up of notes C, E and G (or 0, 4 and 7 in integer notation). Rameau noticed that every individual harmonic of those three notes also happens to be a harmonic of a lower, unsounded note, two octaves below the C. He called this unsounded note the *basse fondamentale* or "fundamental bass"—and realized that it meant there was no real difference between chords made up of any permutation of notes C, E and G. As Pierce explains:

> Rameau was the first to insist that what we now call various "inversions" of a chord are the same chord, because they have the same fundamental bass. That is, E-G-C and G-C-E are the first and second inversions of C-E-G, and we regard them, as Rameau did, as essentially the same chord. Before Rameau, they were named differently and regarded as distinct chords. Rameau thus drastically reduced the number of different chords that a musician must learn [9].

In other words, to adopt the language of mathematics, a chord is an "unordered set" of pitch classes. Using integer notation—which in many contexts is much clearer than musician's letter-codes—a C major chord is (0, 4, 7) regardless of which order the notes are placed in. A set-based approach of this kind can be extremely useful in understanding certain aspects of music, such as the transition that took place in the early 20th century from "tonal" to "atonal" composition. That's what we're going to look at next.

Set Theory

As explained in the first chapter, a MIDI keyboard returns an integer for any key pressed, and an integer difference of one corresponds to the interval musicians refer to as a semitone. There are 12 semitones in an octave, and it turns out that transposing a note up or down by an octave has no effect on harmony. That mean we can use modulo-12 arithmetic, and take our "universe set" as the 12 pitch classes (PCs) 0–11.

Most western music is written in a key comprising a smaller subset of PCs, such as C major which only uses the seven PCs (0, 2, 4, 5, 7, 9, 11). There's a common operation called "transposition", in which all the notes are shifted up or down by a constant number of semitones to put the music in a different key. For example, raising C major by two semitones produces D major. It's unlikely that many musicians think of this as a mathematical operation, but that's exactly what it is, as this quote from Allen Forte, one of pioneers of musical set theory, makes clear:

> The transposition of a PC integer i means that some integer t is added to i to yield a PC integer j. If j is greater than or equal to 12, j is replaced by the remainder of j divided by 12. This is called addition modulo-12, abbreviated to mod-12. From this it is evident that PC set A is equivalent to PC set B if there is some integer t which, added to each integer of A, will yield the corresponding integer in B [10].

Translating that from mathematical to musical jargon, Forte is saying that if you take a composition in C major and raise every note by two semitones you're left with the same composition in D major. His assertion that the two sets are "equivalent" is based on the assumption that the absolute value of a PC is irrelevant, and the only important thing is the interval between adjacent PCs—whether arranged vertically in chords or horizontally in melodies. In this view, C major (0, 2, 4, 5, 7, 9, 11) and D major (2, 4, 6, 7, 9, 11, 1) are essentially identical, since they both have the same intervals between consecutive notes: (2, 2, 1, 2, 2, 2, 1).

Today, most people would agree that it makes little difference whether a tune is sung or played in C major or D major. In the past, however, the view seems to have been very different. If you're a classical music fan, you'll be inured to the fact that virtually every instrumental work written in the 18th or 19th century has a key in the title: "Symphony in G minor", "Piano Concerto in E flat", "String Quartet in C sharp minor" and countless others. If you're not a fan of classical music, perhaps this arcane and alienating tradition is one of the reasons you never got into it.

Nevertheless, it really does seem that people used to be far more sensitive to musical keys—even to the extent of linking different keys to different moods and feelings, as Eli Maor explains:

> Around 1800, keys began to be associated with various emotional attributes. Qualities such as "bright", "heroic" or "tragic" were being liberally use by music critics to characterize different keys, as if the mere designation of a key by name endowed it with emotional powers [11].

The previous chapter described how atonal music became a cliché of horror movies in the 20th century, but in the days when harmonic possibilities were much more limited, the supernatural had to be represented within the existing framework of keys. Referring to Wagner's depiction of the evil sorcerer Klingsor in his opera *Parsifal*, Carolyn Abbate writes:

> Wagner's choice of B minor to represent Klingsor's realm was far from casual. The 18th and 19th centuries had endowed this key with a peculiar effect; for Beethoven it was the *schwarze tonart*, the black tonality. Associations with magic, the supernatural and the malign were a strong part of its character. Within Wagner's works it would be used as an iconic key for the *Flying Dutchman*… It is also the tonal colour given to Alberich's curse in *The Ring* [12].

There's an amusing irony in all of this, because the allocation of musical pitches to specific frequencies has changed over time. As Maor explains, "the modern standard A = 440 Hz was adopted at a congress of physicists in Stuttgart in 1834, but it was not until 1939 that this became the official international benchmark" [13]. He also notes that a tuning fork used by the composer Handel, dating from 1751, is tuned to A = 422.5 Hz, only slightly higher than modern A-flat = 415.3 Hz. As it happens, in that same year of 1751 Handel wrote a "Violin Sonata in A"—perhaps it should be renamed Violin Sonata in A-flat!

The reality, of course, is that our ears are much more sensitive to the intervals between PCs than to their absolute values. In traditional music theory there are 12 different intervals, corresponding to every possible difference between any two PCs. In modern set theory, this is often simplified to just 6 interval classes (ICs). This is achieved by ignoring the octave interval and assuming that, say, the interval of one semitone between the notes C and C-sharp is equivalent to the interval of 11 semitones from that C-sharp up to the next C. The six ICs, with their traditional names in brackets, are then:

- IC1 = 1 or 11 semitones (minor second or major seventh)
- IC2 = 2 or 10 semitones (major second or minor seventh)
- IC3 = 3 or 9 semitones (minor third or major sixth)
- IC4 = 4 or 8 semitones (major third or minor sixth)
- IC5 = 5 or 7 semitones (perfect fourth or perfect fifth)
- IC6 = 6 semitones (tritone)

A particular chord—or any set of notes—is primarily characterized by its interval content. There's an elegant way of expressing this concisely, as Allen Forte explains:

> The description of the total interval content of any PC set is given by its interval vector, which is constructed as follows. Count the number of occurrences of IC1 and write down that number. Count the number of occurrences of IC2 and write that number to the right of the previous number. Continue in this fashion until the number of occurrences of IC6 has been written down. To illustrate, the vector for the set (0, 3, 6, 7) is [1 0 2 1 1 1] [14].

The C major scale (0, 2, 4, 5, 7, 9, 11) has interval vector [2 5 4 3 6 1], and hence includes at least one instance of every IC. On the other hand, the archetypally harmonious C major chord (0, 4, 7) has interval vector [0 0 1 1 1 0], omitting both minor and major seconds and the tritone. The simplest chords that incorporate all 6 ICs are the so-called "all-interval tetrachords", such as [0, 1, 4, 6], which have the elegant-looking (but not necessarily pleasant-sounding) interval vector [1 1 1 1 1 1].

The mod-12 arithmetic of music set theory may seem strange at first, but there's an everyday counterpart in the hours on a clock face. The only difference is that hours run from 1 to 12 while PCs run from 0 to 11. It's an analogy that suggests a neat visual way to represent musical sets, as illustrated in Fig. 4. This shows three different four-note sets:

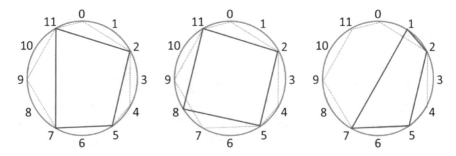

Fig. 4 Circle diagrams for a dominant 7th, a diminished 7th and an all-interval tetra-chord (shown in red), superimposed on the C major scale (dotted blue line)

- a dominant 7th (2, 5, 7, 11), one of the most frequently heard four-note chords
- a diminished 7th (2, 5, 8, 11), with a neatly symmetrical shape
- an example of an all-interval tetrachord (1, 2, 5, 7)

Also shown, as a dotted line, is the basic C major scale (0, 2, 4, 5, 7, 9, 11). While G7 only uses notes from this scale, the other two chords require the addition of another note from outside it.

One thing to look out for in these diagrams is the tritone interval (IC6), which as mentioned earlier in this chapter is a particularly harsh dissonance. It's easy to spot because its two notes, being exactly half an octave apart, are diametrically opposite each other. All three of the examples in Fig. 4 include at least one tritone (there are two of them in the diminished 7th).

For 18th century composers like Mozart, the dominant 7th was the only dissonant chord in common use, and because of the presence of harmonious intervals (IC3 and IC4) in addition to the tritone, it's a very mild dissonance. In Mozart's time, it would have been immediately followed by a more stable chord lacking a tritone—a tonal process referred to as "resolving" a dissonance.

Mozart also used diminished 7ths, but only in exceptional circumstances such as when the statue of a murdered man comes to life in the opera *Don Giovanni*. From a geometric point of view the diminished 7th looks attractive because of its symmetry, but it's this very symmetry that makes the chord problematic for tonally conceived music. It's so symmetrical it isn't obvious which key it belongs to, which means there are many different ways a composer could choose to resolve it (in the Mozart example, he goes to a dominant 7th followed by a minor chord).

Even more dissonant than a tritone is a semitone interval—two adjacent points in the circle diagrams. There's one of these in the all-interval tetrachord (the clue's in the name), but chords like this were much too dissonant for people like Mozart, who was quoted as saying "music, even in the most terrible situations, must never offend the ear" [15].

In later centuries composers became more adventurous in their use of chords—and in their ideas about acceptable levels of dissonance. Despite its colloquial usage, dissonance doesn't automatically mean "unpleasant" or "avant-garde". According to the Classic FM website, an academic survey of contemporary pop songs in 2017 "found that the sound most consistently associated with happy lyrics was the minor 7th chord" [16].

Yet a minor 7th, such as (0, 3, 7, 10), includes a dissonant IC2 interval between PCs 0 and 10. The same IC2 dissonance crops up in the so-called "Tristan chord" (3, 5, 8, 11), which is a recurring feature of Wagner's opera *Tristan and Isolde*. Wagner was famously adored by the Nazis, who were hardly known as patrons of avant-garde music.

As soon as we admit the idea of dissonance into music, we're on the road to atonality—it's just a question of degree. Very few dissonant chords fit into the strict framework of the 7-note tonal scale. The dominant 7th and minor 7th do—but other common dissonances, such as the diminished 7th or Tristan chord, require other notes from outside the scale.

In fact, atonality doesn't even require dissonant chords, since you can have an atonal "melody" where consecutive notes are taken from different keys. Even Mozart dabbled with this idea. The musicologist Hans Keller, writing about one of Mozart's best known works, the 40th symphony (1788), refers to a "shattering moment of threatening disintegration" in the final movement where "ten different notes appear in pretty atonal succession" [17].

A few years later, another composer was faced with a situation that virtually demanded an atonal solution. This was Joseph Haydn, who produced a large-scale choral work based on the Book of Genesis called *The Creation* (1798). The problem came right at the start, with an orchestral prelude which had to depict the state of chaos preceding the creation of the cosmos. While Haydn's solution sounds nothing like as chaotic as a more recent composer would have made it, the music is nevertheless "atonal" in the sense that, for the best part of six minutes, you never really know what key it's in. As Georg Predota writes on the *Interlude* website:

> He starts his portrayal titled "The Representation of Chaos" with a unison C played by the full orchestra. It has no harmony, no dissonance, no melody and no rhythmic pulse … as Donald Francis Tovey wrote, "here is your infinite empty musical space". And from this nothing, Haydn ingeniously assembles the fundamental materials … two notes create an interval, joined by another single

note to create a chord. But this basic triad exists in a harmonic vacuum, as "Haydn denies the existence of tonal organization". Skillfully withholding cadences from the end of phrases, Haydn depicts chaos through progressions that do not adhere to the established conventions of his day [18].

True atonality, however—in the sense of a complete break with the conventions of tonal music—had to wait another hundred years, until the beginning of the 20th century. It's a development that is inextricably linked to the Austrian composer Arnold Schoenberg—who, ironically enough, hated the term "atonal". He put it this way:

> A piece of music will always have to be tonal, at least insofar as a relation has to exist from tone to tone by virtue of which the tones, placed next to or above one another, yield a perceptible continuity. The tonality itself may perhaps be neither perceptible nor provable... Nevertheless, to call any relation of tones "atonal" is just as far-fetched as it would be to designate a relation of colours "aspectral"... If one insists on looking for a name, "polytonal" or "pantonal" could be considered [19].

Nevertheless, in common parlance, "tonal" is reserved for music that follows the kind of rules that composers like Mozart and Wagner used, while "atonal" refers to music that follows different rules—or no rules at all.

It's important to see Schoenberg's music, which continued to evolve throughout his career, in the context of the time it was written. His earliest non-tonal works, dating from just before World War One, reflect the general fear and anxiety of that time, as well as the broader trend in the art of that period towards "expressionism"—valuing emotional content over formal beauty. In Schoenberg's case, these factors make his early atonal works among the most uncompromisingly dissonant music ever composed.

Perhaps the most relevant example in the context of the present book is a 30-minute piece for soprano and orchestra called *Erwartung* ("Expectation"), which Schoenberg wrote in 1909. He called it a "monodrama", and it can either be thought of as a very long song or a very short, one-person opera. Here is how this "angst-dream", as Schoenberg also referred to it, is summarized in the liner notes of a CD recording:

> In the nightmare scenario a lone, anguished woman wanders in a forest, seeking her absent lover, only to discover his dead body. Perhaps she herself has killed him, and returned to the scene in demented forgetfulness. The score—which is shot through with passages of great atmospheric beauty as well as violence—resembles nothing so much as a gigantic written-out improvisation, a musical stream of consciousness [20].

The first thing that hits you about *Erwartung* is the soprano's part—which is, to put it bluntly, shrieky and squawky. It's not an attractive sound, and many people are going to hate it. But if you "listen through" the soprano's hysterics to the underlying orchestral accompaniment, *Erwartung* is an astonishing work. It sounds like the archetypal soundtrack from a 1950s psychological horror film—which is appropriate enough given the narrative, but decades ahead of its time.

As it happens, Schoenberg wasn't too impressed when the style of music he pioneered was assimilated into the soundtracks of movies and popular radio dramas. In 1949 he wrote:

> What displeases many listeners are the dissonances and the absence of a constantly present tonality. It looks as if today's listeners are not enough afraid of such evils and are ready to accept such meaningless noises as the murder and mystery stories of the radio use for background illustration… Today, atonality is tolerated by all radio listeners, on condition that it would not try to say anything sensible [21].

By breaking the established rules of tonality, Schoenberg was able to achieve new levels of expressiveness in works like *Erwartung*. But there's a downside to ditching all the rules of harmony, because those rules serve two distinct purposes. The obvious one, which applies to pop songs as well as symphonies and operas, is a "local" role that governs the way musical notes relate to their immediate neighbours—both vertically in chords and horizontally in melodies. That's something that, in the interests of expressionism, can perhaps be dispensed with.

The second role, which is just as important as the first in large-scale classical works, is a "global" one—to provide a structural framework for the composition. This isn't necessarily obvious to listeners, but it can be very important from a composer's point of view. Schoenberg felt he needed a functional framework of this kind if he was going to compose "real" classical works like symphonies, string quartets and full-length operas.

His solution, which he arrived at in the 1920s, was the so-called "12-tone method"—which was already touched on in the previous chapter due to its use in movies like *The Curse of the Werewolf* and *Planet of the Apes*. But it's worth describing Schoenberg's 12-tone method in a bit more detail at this point—if only because it's a perfect example of the subject of this chapter: musical mathematics.

Enter the Matrix

The traditional tonal system achieves its basic order by focusing on a subset of the 12 pitch classes, such as the major scale (0, 2, 4, 5, 7, 9, 11). Schoenberg's approach was completely different, as one of his followers, the American composer Charles Wuorinen, explains:

> In the 12-tone system, all the available 12 tones are, in principal at least, present all the time, in free circulation. Obviously here, pitch content cannot have a fundamental organizational meaning … something else is needed. It is found in the *order* of pitch and interval occurrence, expressed not as a surface specific, but as a fundamental structural principle. This, then, is the main difference between the tonal and the 12-tone systems: the tonal system is based upon interval content, the 12-tone system upon interval order [22].

More specifically, as Wuorinen goes on to say, "Schoenberg conceived the idea of an abstract source for all the material, melodic and harmonic, in a given work: the 12 pitch classes in a single unique ordering" [22]. This ordered set of PCs is sometimes called a "tone row" and sometimes a "series"—the latter term eventually giving rise to the concept of "serialism", in which Schoenberg's approach is extended beyond pitch classes to other aspects of musical composition.[1]

Schoenberg referred to his new approach as the "method of composing with 12 tones which are related only with one another", which is a far from succinct label—and not a completely accurate one, as his biographer Allen Shawn points out:

> Properly speaking, it is not a "method for composing" but a method for establishing the tonal world of a specific piece out of the 12 notes of the chromatic scale. Once it is established, one must then create a piece of music in this tonal world. This definition makes it plain why there are 12-tone pieces of every level of quality [23].

That's an important point. Anyone can take Schoenberg's rules and use them to compose bad music, just as bad music can be produced by following the traditional rules of tonal music. In a very loose sense, the tone row in Schoenberg's approach fulfils a similar role to the key in traditional music—although it's used in very different ways. The biggest difference comes from the fact that a key is an unordered set of notes, while a tone row is an ordered

---

[1] A discussion of serialism in this broader sense will be deferred to a later chapter, "Scientific Music".

set—and its ordering is of critical importance to Schoenberg's method. Quoting Shawn again:

> He found that if he used a specific succession of notes as his basis for a composition, the ear would hear any other version of the same succession of the intervals created by these tones as belonging to the same tonal family... If a duck is reflected in a pond we recognize the mirror image of the duck without thinking of it as a second duck. In the same way, the inversion of the basic set is simply the reflection of the original set, and the retrograde is simply the set played backwards [23].

The use of terms like "inversion" and "retrograde" brings us to the most important feature of the 12-tone method as far as this book is concerned: its mathematical nature. Earlier in this chapter it was pointed out that the process musicians call "transposition" is basically a mathematical operation, and in Schoenberg's approach it's joined by other similar operations. Here is Charles Wuorinen on the subject:

> Ever since the initial development of the 12-tone system by Schoenberg, there have been certain fundamental transformations to which 12-tone sets are routinely subjected... The basic operations are: transposition, inversion, retrogression. When appropriately combined, these transformations yield a group or array of set-forms all of which are closely related to the original set [24].

Transposition, as defined earlier, means adding a constant number of semitones to every element of the series—modulo-12, of course, as Wuorinen explains:

> If we transpose 0 by 7, 7 results, but if we transpose 7 by 7 again, we reach not "pitch class 14", which doesn't exist, but pitch class 2, which is 14 mod 12 [24].

The two other important operations are retrogression—which simply means reversing the order of the series—and inversion, which Wuorinen defines as follows:

> The inversion of a pitch or interval class is its complement mod 12—that is, it is the difference between the original quantity and 12 [24].

Since there are 12 possible transpositions of a series, and each of these can exist in inverse, retrograde and retrograde-inverse variants as well as the prime form, we end up with 48 different versions of the series. Fortunately, there's a concise way of depicting all 48 of these in the form of a 12 × 12 matrix, as illustrated in Fig. 5.

To start with, the prime series is shown horizontally, from left to right, on the top row of the matrix. If this starts with PC zero, as it does in this example, it is labelled P0. When it's read backwards, from right to left, this row also represents one of the retrograde series—R7 in this case, since the P0 series ends with 7.

	I0	I11	I8	I2	I3	I5	I9	I6	I1	I4	I10	I7	
P0	0	11	8	2	3	5	9	6	1	4	10	7	R7
P1	1	0	9	3	4	6	10	7	2	5	11	8	R8
P4	4	3	0	6	7	9	1	10	5	8	2	11	R11
P10	10	9	6	0	1	3	7	4	11	2	8	5	R5
P9	9	8	5	11	0	2	6	3	10	1	7	4	R4
P7	7	6	3	9	10	0	4	1	8	11	5	2	R2
P3	3	2	11	5	6	8	0	9	4	7	1	10	R10
P6	6	5	2	8	9	11	3	0	7	10	4	1	R1
P11	11	10	7	1	2	4	8	5	0	3	9	6	R6
P8	8	7	4	10	11	1	5	2	9	0	6	3	R3
P2	2	1	10	4	5	7	11	8	3	6	0	9	R9
P5	5	4	1	7	8	10	2	11	6	9	3	0	R0
	RI5	RI4	RI1	RI7	RI8	RI10	RI2	RI11	RI6	RI9	RI3	RI0	

Fig. 5 Example of a 12-tone matrix, showing all 48 variants of the original row in compact form. The 12 transpositions of the prime row are shown horizontally left to right (red labels), the inversions vertically downwards (black labels), the retrogrades horizontally right-to-left (blue labels) and retrograde inversions vertically upwards (green labels)

Next, the inversion I0 is written vertically downwards in the first column, starting from the same zero as the first note of P0. Reading down this first column gives us starting points for all the other prime transpositions (left to right), while the remaining elements of P0 provide starting points for all the other inversions (top to bottom). Reading the prime rows backwards gives us all 12 retrogrades, while reading vertical columns from bottom to top gives the retrograde inversions.

Bearing in mind that we can start or end at any point in the matrix, move through it horizontally or vertically, jump from place to place in it, and use it to generate both chords and melodies, there's obviously a huge amount of musical material here. So it's no surprise that, as Shawn said, an inept composer can make very bad music out of it. At the same time, a great composer can make great music out of it. As Shawn says elsewhere in his book:

> The 12-tone row is not a scale. It is a quarry, in the words of H. H. Stuckenschmidt, from which all the material used in a single work is derived, imparting to it an individual sound world, its own source material in the form of a particular sequence of intervals. It is like a musical DNA chain [25].

Schoenberg himself used his system as skilfully as anyone. Not all his 12-tone works are masterpieces, but some of them are. Possibly the greatest of all is his opera *Moses and Aron*, which is more dramatic, faster moving and a lot more fun to listen to than the aforementioned *Tristan and Isolde* by Wagner.[2] It's also a cleverer and more thought-provoking narrative. Loosely based on the Biblical story of Moses and Aaron[3]—the former receiving the Ten Commandments from God, the latter creating the "Golden Calf" idol—Schoenberg transforms it into a much more subtle psychological drama. Here's how the liner notes of the Naxos recording describe it:

> The position that Aaron assumes as interpreter and communicator between Moses and the people is often dismissed as that of a demagogue and agitator. In fact he is a mediator who shares the thinking of Moses and at the same time knows of the people's need for concrete images ... Aaron's ability as a speaker is the capacity to find metaphors and interpret them. He is in a position to reach the people with imagery and make it possible for them to understand themselves ... Aaron's gift of verbal communication and the divine miraculous signs

---

[2] This is a personal opinion!

[3] If you're confused by the two different spellings—Aron in the opera's title and Aaron everywhere else—hold that thought until the end of this chapter.

that are intended to legitimize the mission of Moses—two quite different themes in the Biblical account—Schoenberg regards as identical [26].

In Schoenberg's opera, there's no black-and-white distinction between the literalist religion of Moses and the metaphorical religion of Aaron—just as, for him, there was no black-and-white distinction between "tonal" and "atonal" approaches to composition:

> The tablets of the law and the Golden Calf: in the radical interpretation of Schoenberg these—the analogy is with the compositional style—are in no way categorically different: they are emanations of one and the same abstract series method, an aesthetic cypher for the inexpressible divine name [26].

Moses and Aron isn't science fiction, but its attempt to rationalize a Biblical story in modern terms is reminiscent of similar exercises by SF authors, such as the reimagining of the Christ narrative in Michael Moorcock's "Behold the Man" (1966).[4]

Although Schoenberg didn't write any explicitly sci-fi music, his one excursion into movie music—for a film that only existed in his head—is so generic that it fits a sci-fi scenario as well as any other. To quote Allen Shawn's biography again:

> One of Schoenberg's most immediately appealing 12-tone works is his concise and moody *Accompaniment to a Film Scene*. Although the film scene was imaginary, Schoenberg appended a useful synopsis of the score: "Fear—Threatening Danger—Catastrophe" [27].

Although the atonal style pioneered by Schoenberg proved well-suited to genres like sci-fi, horror and suspense, it's totally unsuited to the romantic and sentimental themes that infuse the great bulk of music past and present. There is, however, one other area where Schoenberg's method works well—and that's comedy. He produced several playful compositions himself, such the Suite for septet, opus 29—"a piece of essentially light-hearted character", according to musicologist Arnold Whittall, in which "Schoenberg contrives something which sounds like highly intellectual, utterly Teutonic jazz" [28].

Other composers took 12-tone music even deeper into comedic territory. In fact, you probably heard examples of it as a child without realizing it. As Kathryn Kalinak writes in her *Very Short Introduction to Film Music*:

---

[4] Moorcock is a double-hatted writer-musician, as we'll see in a later chapter, "Science Fiction and Music Culture".

Serialism is a method of composition associated with Arnold Schoenberg in which all 12 pitches in the Western scale are used equally to avoid establishing any tonality. Listen for it in Scott Bradley's scores for *Tom and Jerry* cartoons at MGM. Quipped Bradley, "I hope that Dr Schoenberg will forgive me for using his system to produce funny music, but even the boys in the orchestra laughed when we were recording it." [29]

The "cartoony" potential of atonal music also struck the composer John Adams, who mixed Schoenberg with Roadrunner in his *Chamber Symphony* of 1992. Here is Adams himself on the subject:

The *Chamber Symphony* bears a superficially suspicious resemblance to its eponymous predecessor, the opus 9 of Arnold Schoenberg.[5] The choice of instruments is roughly the same as Schoenberg's, although mine includes parts for synthesizer, percussion, trumpet and trombone. However, whereas the Schoenberg symphony is in one uninterrupted structure, mine is broken into three discrete movements, "Mongrel Airs", "Aria with Walking Bass" and "Roadrunner". The titles give a hint of the general ambience of the music.

He goes on to explain the work's genesis:

I was sitting in my studio, studying the score to Schoenberg's *Chamber Symphony*, and as I was doing so I became aware that my seven-year-old son was in the adjacent room watching cartoons. The hyperactive, insistently aggressive and acrobatic scores for the cartoons mixed in my head with the Schoenberg music, itself hyperactive, acrobatic and not a little aggressive, and I realized suddenly how much these two traditions had in common [30].

An equally humorous (if more risqué) piece containing a 12-tone segment is the song "Brown Shoes Don't Make It" by Frank Zappa and the Mothers of Invention, from their 1967 album *Absolutely Free*. The Schoenberg-inspired section is the bit beginning "we see in the back of the City Hall mind the dream of a girl about thirteen" (copyright laws—and probably other laws too—prevent us from quoting any more of the lyrics). Interestingly, the basic 12-tone series used in "Brown Shoes Don't Make It" was originally created by Zappa a decade earlier, when he was still a teenager, for a composition he wrote called "Waltz for Guitar" [31].

---

[5] This dates from 1907, during Schoenberg's early atonal period, and well before he devised his 12-tone method. It's also several decades before the first Hollywood cartoons, but anyone hearing Schoenberg's *Chamber Symphony* today is likely to be reminded of them.

As surprising as it may seem, Zappa was a passionate devotee of avant-garde music throughout his life. In John Corcelli's book *Frank Zappa FAQ* we're told that when a radio show asked Zappa for his top ten records in 1989, his selection included pieces by Edgard Varèse, Igor Stravinsky, Béla Bartók, Maurice Ravel and Anton Webern—the latter, a pupil of Schoenberg, represented by both a string quartet and a symphony. Corcelli adds that, in an interview in 1976, "Zappa said he listened to Webern's quartets when he wanted to relax" [32].

Frank Zappa was never one to tolerate bullshit, or to profess an interest in something simply because it was fashionably avant-garde, so his admiration for a 12-tone composer like Webern is important. If he said Webern wrote great music, it was because Webern wrote great music.

The mathematical aspects of the 12-tone system—the aspects we've been focusing on in this chapter—are really only incidental to the fact that a few of its exponents, such as Schoenberg and Webern, were able to create masterpieces using it. Yet for Schoenberg, the mathematical nature of his method remained important alongside its musical potentialities, as Allen Shawn explains:

> The mathematical and the deeply musical—innovation and tradition—were deeply wedded. All of Schoenberg's tendencies—the need for order, for unity of materials combined with unceasing development... his love of games and reverence for numbers—came together in this approach [23].

At the start of this chapter, we saw that even the most traditional forms of music have a mathematical side, but in most cases it's hidden well below the surface. 12-tone music, on the other hand, is almost impossible to discuss without resorting to mathematical terminology. At a quick glance, even Schoenberg's own notes for *Moses and Aron* (Fig. 6) look more like mathematics than music.

In the end, though, Schoenberg was an artist rather than a mathematician, and his attitude to numbers wasn't an entirely rational one. For example, he had a superstitious fear of the number 13. Among other things, this explains a minor mystery that may have been puzzling some readers. Why does he spell the name of his opera's protagonist "Aron", when the Bible—and everyone else—spells it "Aaron"? Allen Shawn has the answer:

> Schoenberg's relationship to numbers has its sublime and its ridiculous aspects. The number 12, with which his name is forever associated, is one less than the number of which he had a morbid fear. An instance he cited was page 13 of his

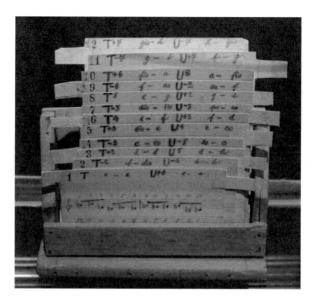

Fig. 6 A set of notes Schoenberg produced for his opera *Moses and Aron* looks, at first glance, more mathematical than musical (public domain image)

Violin Concerto, the point in the score where his work was interrupted by a three-week illness and where he later made a mistake in the numbering of measures. He removed an "a" from the spelling of Aaron's name in *Moses and Aron* so that the title would not contain 13 letters [33].

References

1. I. Asimov, *Biographical Encyclopedia of Science and Technology* (Pan Books, London, 1975), p. 5
2. E. Maor, *Music by the Numbers* (Princeton, Princeton University Press, 2019), p. 15
3. J. Charon, *Cosmology* (Weidenfeld & Nicholson, London, 1970), pp. 21–22
4. D.K. Love, *Kepler and the Universe* (Prometheus Books, New York, 2015), pp. 160–162
5. S.D. Tucker, *Space Oddities* (Amberley, Stroud, 2017), p. 53
6. P. Greco, *Galileo Galilei: The Tuscan Artist* (Springer, Switzerland, 2018), p. 73
7. G. Galilei, *Dialogue concerning Two New Sciences*, http://galileoandeinstein.physics.virginia.edu/tns_draft/tns_061to108.html
8. J.R. Pierce, *The Science of Musical Sound* (Scientific American Books, New York, 1983), pp. 42–44

9. J.R. Pierce, *The Science of Musical Sound* (Scientific American Books, New York, 1983), p. 91

10. A. Forte, *The Structure of Atonal Music* (Yale University Press, New Haven, 1973), p. 6

11. E. Maor, *Music by the Numbers* (Princeton, Princeton University Press, 2019), pp. 109–110

12. C. Abbate, Parsifal: Words and Music, in *Parsifal*, (English National Opera, London, 1982), p. 51

13. E. Maor, *Music by the Numbers* (Princeton, Princeton University Press, 2019), pp. 82–83

14. A. Forte, *The Structure of Atonal Music* (Yale University Press, New Haven, 1973), p. 15

15. E. Maor, *Music by the Numbers* (Princeton, Princeton University Press, 2019), p. 10

16. "Happy Chords Are Dying Out, Study Reveals", Classic FM, November 2017., https://www.classicfm.com/music-news/happy-chords/

17. R. Simpson (ed.), *The Symphony*, vol vol. 1 (Penguin, London, 1966), p. 96

18. G. Predota, What Does Nothing Sound Like in Music?, *Interlude*, November 2018.,   https://interlude.hk/nothing-sound-like-music-haydns-representation-chaos/

19. K. Gloag, N. Jones, *The Cambridge Companion to Michael Tippett* (Cambridge University Press, Cambridge, 2013), p. 8

20. M. MacDonald, *Schoenberg Orchestral Works* (CD liner notes, Warner, 2008)

21. F. Sherry, *Schoenberg String Quartets Nos 3 and 4* (CD liner notes, Naxos, 2010)

22. C. Wuorinen, *Simple Composition* (Longman, New our, 1979), pp. 5–6

23. A. Shawn, *Arnold Schoenberg's Journey* (Harvard University Press, Massachusetts, 2002), pp. 196–199

24. C. Wuorinen, *Simple Composition* (Longman, New Your, 1979), pp. 85–88

25. A. Shawn, *Arnold Schoenberg's Journey* (Harvard University Press, Massachusetts, 2002), p. 205

26. S. Morabito, *Moses and Aron* (CD liner notes, Naxos, 2006)

27. A. Shawn, *Arnold Schoenberg's Journey* (Harvard University Press, Massachusetts, 2002), p. 221

28. A. Whittall, *Schoenberg: Suite, opus 29* (CD liner notes, Decca, 1992)

29. K. Kalinak, *Film Music: A Very Short Introduction* (Oxford University Press, 2010), p. 69

30. J. Adams, *Chamber Symphony* (CD liner notes., Elektra, 1994)

31. K. Sloots, Absolutely Free: Complexities. *Zappa Analysis*, https://www.zappa-analysis.com/absolutely-free.htm

32. J. Corcelli, *Frank Zappa FAQ* (Backbeat Books, Milwaukee, 2016), p. 15, 23

33. A. Shawn, *Arnold Schoenberg's Journey* (Harvard University Press, Massachusetts, 2002), p. 148

The Electronic Revolution

Electronic music has come a long way since the days when it was associated almost exclusively with science fiction in people's minds. In this chapter we trace the early history of the genre, from early experiments with tape recorders and *musique concrète*, through the laboratory-based work of composers like Karlheinz Stockhausen and Pierre Boulez—not to mention the BBC's Radiophonic Workshop, which was responsible for the *Dr Who* theme tune—to the use of computers in "algorithmic composition". After that we see electronic music entering the mainstream, with the work of pioneering artists like Brian Eno, Tangerine Dream and Wendy Carlos, who produced the soundtrack for Stanley Kubrick's *A Clockwork Orange*.

Composers in the Laboratory

No musical form has undergone such a complete reversal of reputation as electronic music. Today it's the cheapest and easiest kind of music to produce, with most people owning a smartphone, tablet or laptop computer capable of running sophisticated music creation apps like Apple's GarageBand. As a consequence, electronic music has become ubiquitous, associated more than anything else with popular dance music. But 60 or 70 years ago the situation couldn't have been more different. Producing electronic music was a slow and difficult process, requiring custom-built, specialist equipment, and the only people who ventured into the field were serious composers of avant-garde classical music.

© The Editor(s) (if applicable) and The Author(s),
under exclusive licence to Springer Nature Switzerland AG 2020
A. May, *The Science of Sci-Fi Music*, Science and Fiction,
https://doi.org/10.1007/978-3-030-47833-9_3

The attraction of electronics for today's musicians is its ease of production—but what was its appeal in the days when it was difficult to create? The answer lies in the vast range of new sound textures—or "timbres", to use musician's jargon—opened up by electronic technology. There are basically two different ways these can be produced. On the one hand, electronic circuits can be used to generate new sounds from scratch, totally unrelated to any acoustic sound source. Examples of this approach—in the form of electronic instruments like the theremin and the bespoke circuits used by Bebe and Louis Barron for their groundbreaking *Forbidden Planet* soundtrack—were encountered in the first chapter, "Alien Sounds".

The second approach was also touched on in the same chapter, in the context of John Cage's *Williams Mix* (1953). As described there, this was created by splicing together fragments of magnetic tape on which the Barrons had recorded hundreds of brief snatches of musical and non-musical sounds. If the circuit-based approach can be seen as the distant ancestor of today's electronic synthesizers, then this tape-based approach is the progenitor of modern "sampling" technology. Under the eyecatching if peculiar name of *musique concrète*, it first emerged in the late 1940s with the wide availability of commercial tape recorders (Fig. 1).

Musique concrète was the brainchild of Pierre Schaeffer (1910–95), a French radio engineer both of whose parents were musicians, giving him the perfect grounding to become a pioneer of electronic music. In 1948 he produced *Etude aux Chemins de fer* (1948)—"a three-minute piece made by

Fig. 1 The advent of magnetic tape recorders, such as this example from the 1940s, opened up a whole new range of musical possibilities (public domain image)

manipulating recordings of railway trains", according to musicologist Paul Griffiths. Although simple novelty records had been made along these lines in the past, Schaeffer's work opened up vast new musical possibilities, as Griffiths goes on to explain:

> It remained to Schaeffer to discover and use the basic techniques of sound trans-formation: reversing a sound by playing its recording backwards, altering it in pitch, speed and timbre by changing the velocity of playback, isolating elements from it, and superimposing one sound on another [1].

From a sci-fi point of view, probably the most noteworthy exponent of *musique concrète* was Desmond Leslie (1921–2001). As mentioned in a previous book in this series, *Pseudoscience and Science Fiction*, Leslie was one of the first authors to produce extravagant speculations on the subject of UFOs—or "flying saucers", as they were originally dubbed—and also dabbled in science fiction, writing the screenplay for the B-movie *Stranger from Venus* (1954).

Even before Leslie branched out into the world of music, the subject had cropped up in his ufological writings. Among his far-fetched assertions was the idea that UFOs were related to the "vimanas", or flying chariots, described in ancient Indian texts. The musical connection came from the texts themselves, which Leslie quotes as telling us that "a vimana can be moved by tunes and rhythms" and that "by music alone some were propelled" [2].

He goes on to cite the following incident, which took place in Switzerland in 1950, as evidence that the same is true of UFOs:

> Many people, including a professor of physics, reported 80 to 100 flying saucers passing overhead. "As they passed they made a noise like an organ," said the professor. Others described it as the sound of a tremendous chord of music—a "celestial symphony". And on 22 May 1947 clusters of flying saucers shot across Denmark at tree-top height making a deep, tuneful humming sound [2].

As silly and unscientific as they look today, Leslie's writings on UFOs were nevertheless culturally significant in helping to create the popular myth of flying saucers as envisioned in the 1950s. Similarly, while his experiments in *musique concrète* are hardly great music, they did their bit in consolidating the relationship between weird electronic sounds and science fiction. Produced between 1955 and 1959 for various films and TV and radio dramas, he later compiled the pieces into an album called *Music of the Future*, which was reissued on CD in 2005. Here is a quote from Leslie taken from the back cover of the CD:

As I see it, *musique concrète* is the arrangement and selection of sound patterns into an intelligent, evocative and potent new musical form. Its basic instrument is the magnetic tape recorder so that the composer has the great advantage of becoming orchestra and conductor as well... The tape recorder, coupled with other devices, can produce an almost unlimited variety of sound spectra. Hence the first problem is not the creation of new sounds but to select and organize [3].

The science-fictional flavour of the resulting sound can be gathered from Leslie's notes for one of the pieces, called *Music from the Voids of Outer Space*:

Opening with "Asteroids", we appear to journey through, and leave behind, that ruined part of the Solar System known as the Asteroid Belt. We are told musically of the loneliness and desolation of this great ring of cosmic debris and fractured bodies that lies between the orbits of Jupiter and Mars... The belt is either the ruins of a world that got too smart and blew itself up or the embryo beginnings of a future globe. In any case it is not a pleasant place [3].

While the use of a tape recorder may have been high-tech for its time, the production of raw sounds to go on the tape was often distinctly low tech—as Leslie makes clear in his description of another section of the same work, depicting the planet Mercury: "I have only used two sound sources, a humming top and a motor horn employing a single, rhythmic pattern throughout". As for another piece on the CD—music for the film *The Day The Sky Fell In*, which Leslie says "caused a near riot at the Venice Film Festival in 1959"—he writes that: "the opening blasts of the play-in started life as a fast electric fan chucked into the strings of a grand piano".

A notable thing about both Pierre Schaeffer and Desmond Leslie is that they came from outside the musical establishment, with expertise primarily in other fields. As a general rule, established composers didn't take developments like *musique concrète* very seriously. The most high-profile exception was Edgard Varèse—who, as a young man at the start of the 20th century, was told by the great French composer Claude Debussy "you have a right to compose what you want to, and the way you want to" [4].

More than half a century later, at the age of 75 in 1958, Varèse produced a remarkable work called *Poème Électronique*, which his protégé Chou Wen-Chung describes in the following way:

Using tape alone, this was played through more than 400 loudspeakers inside the Philips Pavilion designed by Le Corbusier for the 1958 Brussels World Fair. Varèse worked at the Philips laboratories in Eindhoven to produce the montage of unmodified sounds: machine noises, bells, piano, percussion and pure electronic sounds. The effect of these sounds slowly turning and colliding in the continuous 360-degree sound space must have been overwhelming [4].

Even if they weren't as explicitly science fictional as Desmond Leslie's compositions, Varèse often gave his works sciencey sounding titles, such as *Hyperprism* (1923), *Ionisation* (1931) and *Density 21.5* (1936). The same is true of the best known and most influential electronic musician of all, Karlheinz Stockhausen (1928–2007)—albeit, in his case, in his native German. Among his numerous works can be found *Formel* (formula), *Kurzwellen* (shortwaves), *Tierkreis* (zodiac), *Strahlen* (rays) and *Weltraum* (outer space).

With Stockhausen we've moved away from *musique concrète* and "sampling" to the other main approach to electronic music—sound synthesis. The essential principle here comes from the concept of Fourier analysis, as discussed in the previous chapter, "Musical Mathematics". Just as any acoustically produced sound can be decomposed into a series of sine waves, so a number of pure sine tones can be added together to "synthesize" a more complex sound (see Fig. 2).

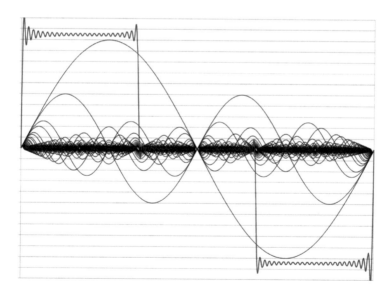

Fig. 2 Any desired sound signal can be synthesized by adding together sine waves of different frequencies and amplitudes—as shown by the black lines in this example, where they yield an approximation to a square wave, shown by the red line (public domain image)

Quoting musicologist Paul Griffiths again:The work of Helmholtz and Fourier had suggested that any sound could be analysed as a collection of pure frequencies, of sine tones, and this was something that Stockhausen thought he had confirmed, in analysing instrumental sounds in Paris. So it seemed reasonable to suppose that the process could be reversed, that timbres could be synthesized by playing together a chosen group of sine tones at chosen relative dynamic levels. One could thereby form a repertory of artificial timbres related in defined ways.

That's exactly what Stockhausen proceeded to do, as Griffiths explains:He found the equipment that enabled him to make the first sine-tone composition, his *Studie I* (1953), in which each sound is constructed from up to six pure frequencies [5].

As commonplace as musical synthesis is today, it's easy to underestimate how revolutionary this was at the time. Up to this point, equipment like sine-tone generators had only ever been used in scientific laboratories. As Stockhausen himself later recalled:

I started looking in acoustic laboratories for sources of the simplest forms of sound wave, for example sine wave generators, which are used for measurement. And I started very primitively to synthesize individual sounds by superimposing sine waves in harmonic spectra, in order to make sounds like vowels, *aaah, oooh, eeeh* etc, then gradually I found how to use white noise generators and electric filters to produce coloured noise, like consonants: *ssss, sssh, ffffh* etc, and when I pulsed them it sounded like water dripping [6].

Creating new musical timbres using laboratory equipment designed for scientific tasks was hard enough, but it was only the first step in the process. Putting those sounds together to create a full-length musical work, such as Stockhausen's 35-minute composition *Kontakte* from 1960, was an even slower process—as he makes painfully clear in the following anecdote:

In some sections of *Kontakte* I had to splice it all by hand, which is an unbelievable labour. Imagine, I worked on the last section of *Kontakte* together with Gottfried Koenig in studio 11 on the third floor of Cologne Radio for three months. And when it was completely ready, I spliced it together with the previous sections, listened, turned pale, left the studio and was totally depressed for a whole day. And I came back next morning and announced to Koenig that we had to do it all over again. I mean, he almost fainted [7].

Part of the problem was of Stockhausen's own creation—a case of "more haste, less speed". Like other exponents of electronic music in the early days, he was keen to rush ahead and create new types of music, which he did in an *ad hoc* and inefficient way using whatever equipment he could lay his hands on (Fig. 3). As another experimental musician of that period, Pierre Boulez, recalled:

> Apparatus intended in theory for measuring or other purposes was diverted from its true function to serve in the creation of do-it-yourself music. As a result, the time it took the first composers in the field to write extremely short compositions (the yield would be a few seconds of music for 15 or even 30 hours of studio time) annoyed everyone and people lost patience. If he had 30 hours of studio time in front of him the composer wanted to produce something of appropriate solidity [8].

A more considered approach would have been, before even touching a piece of electronic apparatus, to think through exactly what you wanted to do, and then build or commission a properly designed suite of equipment to do that precise task. One of the first organizations to take this more business-like approach was the BBC, which set up its Radiophonic Workshop in 1958.

Fig. 3 Karlheinz Stockhausen, pictured in 1994 in the Studio for Electronic Music in Cologne (Kathinka Pasveer, CC-BY-SA-3.0)

Originally intended to provide background music and sound effects for TV and radio shows, the Radiophonic Workshop acquired a much higher profile in 1963 when it created perhaps the most enduring electronic composition of all—the theme music for the sci-fi series *Dr Who*. As described by David Howe, Mark Stammers and Stephen Walker in their history of the show, its original producer Verity Lambert knew she wanted an avant-garde sound for the theme music, but the electronic dimension was an almost accidental afterthought:

> Another essential component was the series' film music. Verity Lambert's first idea was that this should be commissioned from an avant-garde French group called *Les Structures Sonores*... However head of TV music Lionel Salter pointed her instead to the BBC's own Radiophonic Workshop. She visited its Maida Vale studios and spoke with its head, Desmond Briscoe, explaining that what she was looking for was something radiophonic with a strong beat which would sound "familiar but different" [9].

The task of realizing this vision was assigned to Radiophonic Workshop composer Delia Derbyshire:

> She and her assistant Dick Mills used sine and square-wave generators, a white-noise generator and a special beat frequency generator. The tune was put together virtually note by note—each swoop in the music was a carefully timed hand adjustment of the oscillators—and the sounds were cut, shaped, filtered and manipulated in various ways until the tracks were ready for mixing and synchronization [9].

Computers and Music

In an age when "electronics" almost always means "digital electronics", it can be difficult to remember there was ever such a thing as "analog"—but that's what we've been talking about so far in this chapter. A sound is essentially an acoustic waveform, and if the voltage in an electronic circuit is made to fluctuate up and down in exactly the same pattern as that waveform, that's what is meant by an analog representation of it. In a digital representation, on the other hand, the waveform is first converted to a string of numbers, and it's this string of numbers that is stored and manipulated electronically.

Computers are the archetypal digital machines, and because they're programmable they're capable of a whole range of tasks. They can sample or synthesize musical sounds as effectively—and much more quickly—than analog equipment such as tape recorders and sine-wave generators. On top of that, they can do other things that are unique to the digital world, such as stringing sounds together to make a musical composition that obeys whatever logical rules the computer program was given.

This was, in fact, one of the first ways computers were used in music—and it's an inevitable consequence of the trend towards mathematical formulations of music that was described in the previous chapter. As Paul Griffiths explains:

> A great deal of music since 1945 seems, if perhaps only in retrospect, to have been leaning towards cybernetics: the idea of music as sounding numbers, the importance of rules and algorithms in composition, the development of electronic sound synthesis... Images of the composing mind during this period, as evidenced not only in music but in writing about music, tend to suppose rational selection and combination rather than inspiration. Music created with computers is, therefore, part of a much wider concurrence of music and computing [10].

Griffiths wrote that in 1995, and some of his less scientifically savvy readers at the time may have been puzzled by his use of the word "algorithms". Now, thanks to "search engine algorithms" and "social media algorithms", most people will have at least a vague idea what the word means. In effect an algorithm is a set of rules, laid out in a way that is clear and unambiguous enough for a machine to follow, that explain how to achieve a desired result.

One of the first composers to make explicit use of algorithms was Iannis Xenakis, who—perhaps significantly—worked as an engineer before he turned to music. Among other things he devised a form of composition that he called "stochastic music", which we'll learn more about in the next chapter, "Scientific Music". For now, it's sufficient just to quote another snippet from Griffiths's book:

> Xenakis used a computer to handle the manifold calculations that his stochastic music had previously required him to make by hand; an early example is his ST4–1,080262 (1955–62) for string quartet [10].

The novelty here is in the use of computers, not the use of algorithms. Virtually every piece of music ever written is algorithmic at some level, in the sense that it follows clearly defined rules. This fact was recognized centuries ago, as the following quote from Wikipedia makes clear:

A *Musikalisches Würfelspiel* (German for "musical dice game") was a system for using dice to randomly generate music from precomposed options. These games were quite popular throughout western Europe in the 18th century. Several different games were devised, some that did not require dice, but merely choosing a random number… Examples by well known composers include C. P. E. Bach's *A Method for Making Six Bars of Double Counterpoint at the Octave Without Knowing the Rules* (1758) [11].

Even Mozart dabbled in algorithmic composition—allegedly. In 1793, two years after the composer died, his publisher put out a *Musikalisches Würfelspiel* in his name (see Fig. 4). It came complete with directions in four languages, including rather poor English: "Instruction to compose without knowledge of music so much German Waltzes as one pleases, by throwing a certain number with two dice" [12].

Dice games aside, the idea of fully automated composition using computers goes all the way back to the person commonly recognized as the world's first "computer programmer", the Victorian mathematician Ada Lovelace. In those days computers were mechanical contraptions, not electronic ones—and they tended to exist more in imagination than reality—but the basic principles of programming were the same as today. As Lovelace wrote in 1843, in the context of one hypothetical design, Charles Babbage's "Analytical Engine":

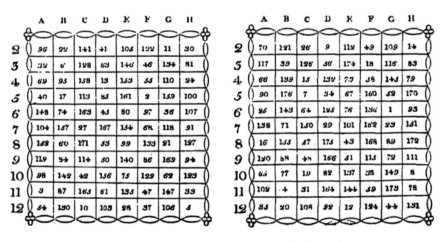

Fig. 4 Two tables from Mozart's *Musikalisches Würfelspiel*, enabling the user to choose appropriate fragments of music by throwing a pair of dice and then reading off the appropriate index numbers from the tables (public domain image)

The Analytical Engine is not merely adapted for tabulating the results of one particular function and of no other, but for developing and tabulating any function whatever... It might act upon other things besides number, were objects found whose mutual fundamental relations could be expressed by those of the abstract science of operations, and which should be also susceptible of adaptations to the action of the operating notation and mechanism of the engine. Supposing, for instance, that the fundamental relations of pitched sounds in the science of harmony and of musical composition were susceptible of such expression and adaptations, the engine might compose elaborate and scientific pieces of music of any degree of complexity or extent [13].

Fast-forwarding a hundred years from the middle of the 19th to the middle of the 20th century, we come to another important figure in the history of computer music, John R. Pierce (Fig. 5). An engineer working at Bell Laboratories in New Jersey, he's already been quoted several times in the previous two chapters. He also featured quite prominently in a previous book in this series, *Rockets and Ray Guns: The Sci-Fi Science of the Cold War*.

As mentioned there, Pierce was the author of around 20 pseudonymous non-fiction articles that were published in *Astounding Science Fiction* in the 1940s and 50s under the pen name of J. J. Coupling. One of these, "Science for Art's Sake" from November 1950, even refers to the subject of algorithmic

Fig. 5 John R. Pierce (1910–2002)—electronic engineer, science fiction author and pioneer of computer music (NASA image)

composition. But this was in Pierce's pre-computer days, and it uses the same time-honoured approach as Mozart: "by the throwing of three especially made dice and by the use of a table of random numbers, one chord was chosen to follow another" [14].

Pierce was a prolific writer on a range of levels, all the way from science fiction short stories such as the much-reprinted "Invariant" (1944) up to serious engineering textbooks. Among the latter is *The Science of Musical Sound* (1983), which has already been quoted several times. Here's what the book has to say on a subject Pierce himself contributed a great deal to, computer music:

> How is it possible for a computer to generate sounds? The sampling theorem gives us a clue. Consider any waveform made up of frequency components whose frequencies are less than B. That is, consider any sound wave whose frequency components lie in the bandwidth between zero and B. Any such waveform or sound wave can be represented exactly by the amplitudes of 2B samples per second. These samples are merely the amplitudes of the waveform at sampling times spaced 1/2B apart in time [15].

The use of the word "sampling" is a little confusing, because it has a different meaning for musicians and engineers. In music, sampling means using a small snippet of a pre-existing audio recording as a component in a new composition. To an electronic engineer, however, sampling means taking an instantaneous measurement of the sound level—just a single number per sound channel. This is done a large number of times per second—2B times in Pierce's terminology—to ensure a faithful representation of the original sound. For a CD-quality digital recording, the bandwidth is B = 22.05 kHz, meaning that 44,100 samples per second are required.

The unique thing about computers, as already mentioned, is their flexibility. As soon as we raise the possibility of representing music in digital form, there's virtually no limit to what can be done with it. Quoting Pierce again:

> Suppose that a composer does have access to a good computer. What does he do with it? If he's a good composer, he doesn't write pieces like those that have been written for orchestra, making them louder, softer, more varied, and "bigger" than he could with conventional instruments. In principle, he could, but in my experience he doesn't. In using new means to produce new sounds, most good composers for the computer want to make a new kind of music [16].

As soon as they had the technology to do so, Pierce and his engineering colleagues began to experiment with creating this "new kind of music". Whether their work had any aesthetic value is questionable, but that wasn't their aim. What they really wanted to do was whet the appetite of professional composers. "After we had produced a small collection of computer-played pieces," Pierce writes, "the Bell Laboratories public relations department had a 10-inch record made, called *Music from Mathematics*".

Early in 1961, copies of the record were sent to two of the most famous composers in America, Leonard Bernstein and Aaron Copland. There was no reply from the former, but the latter showed genuine interest—if only as a "spectator":

> Dear Mr Pierce: It was good of you to send me the *Music from Mathematics* recording. The implication are dizzying and, if I were 20, I would be really concerned at the variety of possibilities suggested. As it is, I plan to be an interested bystander, waiting to see what will happen next [17].

What did "happen next" didn't happen right away. It would be several decades before computer technology was mature enough—and affordable enough—to play a significant role in the music world. By that time, electronic music—in the form of commercially available analog synthesizers—was well and truly established as an indispensable part of popular music.

Into the Mainstream

As it happens, the first mass-produced electronic instrument to make a significant impact on rock music wasn't a synthesizer but a sampler, called the Mellotron. From the outside it looked like an ordinary keyboard instrument, but the interior was like a miniature *musique concrète* laboratory. Functionally, it produced an effect similar to a modern digital sampler, but in an entirely mechanical way, as this description from Wikipedia makes clear:

> Pressing a key causes two screws to connect a pressure pad with the tape head, and the pinch wheel with the continuously rotating capstan. Tape is pulled at a gradual speed, counterbalanced by a tension spring and stored temporarily in a storage bin until the key is released [18].

Although the mechanics of the Mellotron may come across as clunkily Victorian, the unearthly timbres that came out of it were like nothing that had been heard in pop music before. The characteristic Mellotron sound found its way into countless 1960s classics, from the Beatles' "Strawberry Fields Forever" to David Bowie's "Space Oddity".

As for synthesizers themselves—which literally synthesize musical tones from scratch rather than recycling sampled sounds—the first instrument to bear that name was built by RCA in 1955 at the Columbia-Princeton Electronic Music Centre in New York. It was a monstrous contraption, essentially duplicating an entire Stockhausen-style acoustic laboratory in a single package. As Paul Griffiths puts it, "oscillators and noise generators provide the raw materials which the composer, giving the synthesizer its instructions on a punched paper roll, can obtain at will" [19].

The RCA synthesizer was very large, far from cheap and extremely difficult to use—hardly the thing to make chart-topping music with. Its most famous user was as far from the world of rock as you can get—the experimental composer Milton Babbitt, who we'll encounter again in the next chapter, "Scientific Music". Here is how he described the machine many years later:

> That word "synthesizer" connotes some little boy with a small box, sitting at a keyboard. That was far from the case. This was a programmed instrument that was more than the length and size of this room, so people saw it and they thought it was a computer, although it wasn't. It couldn't compute anything, did no number-crunching, and had no memory—for which it was probably grateful... You had to do everything yourself; it was very hard. You programmed every aspect of a musical event and the mode of progression to the next event [20].

The situation changed radically in 1965, when the American engineer Robert Moog (Fig. 6) developed a much more compact synthesizer. This proved practical enough from an ordinary musician's point of view that it could be marketed commercially.

The Doors' 1967 album *Strange Days* was "one of the first uses of the Moog synthesizer in rock", according to the band's keyboardist Ray Manzarek. His description of its use emphasizes how innovative and "technological" the instrument was, in comparison to anything that had gone before: "We were making a record, making music, but we were also mad scientists manipulating the aural spectrum for our diabolical creations" [21].

At that time, the synthesizer was so new that only a few people had mastered it, and the Doors drafted in specialist Paul Beaver to play it on *Strange Days*, rather than Manzarek himself. As the latter recalled later:

Fig. 6 Robert Moog with some of his pioneering synthesizers, including the Sonic 6, Modular 55 and Minimoog (public domain image)

Paul Beaver brought his huge modular Moog system into the studio and began plugging a bewildering array of patch cords into the equally bewildering panels of each module. He'd hit the keyboard and outer space, bizarre, Karlheinz Stockhausen-like sounds would emerge. He would then turn a mystifying array of knobs placed in rows around the patch cord receptacles and more and different space sounds would emerge [21].

Before long, the synthesizer's ability to produce what Manzarek calls "outer space sounds" was snapped up by bands like Hawkwind, who specialized in sci-fi-themed music—a subject we'll take a closer look at in a later chapter ("Science Fiction and Music Culture").

The advent of the synthesizer also helped to blur the boundary between rock and avant-garde classical music. A case in point was the German band Tangerine Dream, as can be seen from this 1974 *New York Times* article on its founder Edgar Froese:

Mr Froese has eliminated the rhythm section from his group entirely, and plays long, instrumental improvisations using three organs, an electric piano, a Mellotron and five synthesizers. And, he says, the band doesn't really care about compromising its standards to cater to its audiences… He cites György Ligeti, Karlheinz Stockhausen, Pierre Henri, John Cage and Terry Riley as his principal influences now [22].

The reference to Ligeti is particularly interesting, given his association with the movie *2001: A Space Odyssey*—which, as we saw in the first chapter, made prominent use of several of his compositions. Ligeti's influence on Tangerine Dream is obvious on albums like *Phaedra* (1974) and *Rubycon* (1975), some passages from which could easily be mistaken for Ligeti's *Requiem* or *Atmosphères*.

Another synthesizer wizard who was stretching the boundaries of rock music around the same time was Brian Eno. His *Discreet Music* of 1975 imports another innovation that had previously only been found in the avant-garde world of people like Stockhausen and Xenakis, namely algorithmic composition. This suited his musical personality very well, as he explains in the liner notes:

> Since I have always preferred making plans to executing them, I have gravitated towards situations and systems that, once set into operation, could create music with little or no intervention on my part [23].

Discreet Music was created with a hybrid analog-digital system—purely digital systems still lying several years in the future. The basic synthesizer used was an analog one, the Synthi AKS from the British company EMS. Uniquely for its time, however, this incorporated what Eno describes as a "digital recall system". In modern terminology this was essentially an audio sequencer, allowing the automated playing of melodic sequences.

For *Discreet Music*, Eno added a further layer of analog processing via a tape-loop feedback system. He considered the resulting configuration such an indispensable part of the composition that an engineering-style "operational diagram" of it is included on the back cover of the album.

As the name suggests, *Discreet Music* was intended from the start to be unobtrusive background music—with no particular associations, science-fictional or otherwise. Nevertheless, the album did acquire a loose connection with sci-fi in the end. As we'll discover in the final chapter, "Speculations on a Musical Theme", *Discreet Music* happened to be a favourite of Philip K. Dick, who modelled one of the characters in his novel *VALIS* after Brian Eno.

As far as science fiction is concerned, however, perhaps the most important early application of synthesizer music is its use in Stanley Kubrick's film *A Clockwork Orange* (1971). In contrast to his previous excursion into sci-fi, *2001: A Space Odyssey*, this provides a very different—and much more pessimistic—vision of the future.

The plot of *A Clockwork Orange* centres around its young anti-hero, Alex, who belongs to a criminal gang devoted to what they call "ultra-violence". It just so happens that (as unlikely as it sounds) Alex is a great fan of Beethoven's music—a fact that becomes highly significant later on, after the authorities catch up with him. When they do, he's subjected to a drastic course of aversion therapy known as the "Ludovico treatment" (see Fig. 7). In the words of film critic Alexander Walker:

> The serum makes Alex unduly receptive to the feeding back into his own system of all the violence in the outside world until he has had such a surfeit of murder, rape and atrocity that his pleasure is turned into revulsion. Significantly, the raw material for the conditioning comes from films… He is a captive audience of one; his head is clamped tight, music is pumped into his ears, a ring of electrodes circles his head like a crown of thorns, and, worst of all, his bulging eyeballs are held forcibly open by surgical lidlocks [24].

The reference to music being pumped into his ears is important, because part of the soundtrack includes Beethoven's Ninth Symphony. As a result, Alex is inadvertently conditioned against this particular piece of music, along with the scenes of violence that it accompanies. Later he naively explains the situation, with foreseeably disastrous consequences, to one of his former victims:

> I'm very fond of music, especially Beethoven—and it just so happened that while they were showing me a particularly bad film of, like, a concentration camp, the background music was playing Beethoven… I can't listen to the

Fig. 7 In Stanley Kubrick's film *A Clockwork Orange*—one of the first to use a synthesizer soundtrack—music plays a significant role in the aversion therapy the anti-hero is subjected to (public domain image)

Ninth any more. When I hear the Ninth, I get like this funny feeling and then all I can think of is trying to snuff it [25].

Needless to say, at the earliest opportunity thereafter, Beethoven's Ninth is duly played in Alex's presence with the aim of causing him to kill himself. As it happens, however, he only succeeds in badly injuring himself.

In the present context, the interesting thing is that the music heard in the film isn't a standard recording of the symphony, but an arrangement of it for synthesizer. It's the work of Wendy Carlos—known at that time as Walter Carlos—about whom the liner notes from the soundtrack CD say the following:

Since his early youth, Carlos displayed a strong interest in both music and scientific technology. When he was 17 years old he assembled an electronic music studio and produced his first electronic compositions which utilized sounds created and manipulated on tape recorders… Intending to develop an electronic sound producer which could validly be termed a musical instrument, Carlos began a collaboration with engineer Robert Moog in 1966. The result was a prototype of Carlos's special synthesizer on which he performed and recorded his realizations of Bach and other composers and his music for *A Clockwork Orange* [26].

Carlos first came to public attention in 1968 with the album *Switched-on Bach*, which consisted of synthesizer arrangements of classical pieces by J. S. Bach—and, impressively enough for a classically inspired work, reached number 10 in the US album charts. A similar treatment applied to Beethoven's music resulted in the version of the Ninth Symphony heard in *A Clockwork Orange*.

After its pioneering appearance in *A Clockwork Orange*, synthesizer music became an increasingly common feature of movie soundtracks—though sometimes in a way that was so close to traditional orchestral music it was difficult to tell the difference. It's a situation that's already been mentioned, in the context of Vangelis's score for *Blade Runner* (1982), in the first chapter. As discussed there, the ambiguity of the music can be interpreted as a reflection of the real-versus-artificial theme of the film itself.

Another sci-fi movie of the same era is David Cronenberg's *Videodrome* (1983), with a synthesizer-rich soundtrack by Howard Shore. As with *Blade Runner*, the film's plot involves artificial constructs impinging on the real world—or, as the back cover of the DVD puts it, "a pulsating science fiction nightmare about a world where video can control and alter human life" [27].

The score of *Videodrome*, like that of *Blade Runner*, helps to underline the blurred distinction between real and unreal. To quote Wikipedia:

> Shore used dramatic orchestral music that increasingly incorporated, and eventually emphasized, electronic instrumentation … to follow the protagonist's descent into video hallucinations. In order to achieve this, Shore composed the entire score for an orchestra before programming it into a Synclavier II digital synthesizer. The rendered score, taken from the Synclavier II, was then recorded being played in tandem with a small string section. The resulting sound was a subtle blend that often made it difficult to tell which sounds were real and which were synthesized [28].

The Synclavier II was one of the first fully digital keyboard instruments, combining both synthesizer and sampler, which came onto the market in 1980. Alongside Howard Shore, another keen user of it was Frank Zappa, who has been mentioned several times in the context of his interest in cutting-edge avant-garde music. His purchase of a Synclavier II in 1982 opened up a whole new area of experimentation, with his biographer John Corcelli quoting him as saying that it "allows me to create and record a type of music that is impossible or too boring for human beings to play" [29].

Corcelli goes on:

> As his music grew in complexity, the Synclavier became an important and invaluable tool, and it remained so for the rest of his life. Finally he could sample the sounds he wanted, add complicated rhythms, play them back, and, if he wanted to, have the music printed, because the system also included a software program known as SCRIPT. We now take this kind of thing for granted, since the advent of Apple's GarageBand and the dozens of other software applications that allow the user to record, mix, edit and compose music, but in 1982 it was adventurous technology for an equally adventurous composer [29].

The first album to feature Zappa's use of the Synclavier was a collaboration with a classical composer who was mentioned earlier in this chapter, Pierre Boulez—albeit in this case in his alternate role as an orchestral conductor. As Corcelli recounts:

> One of Frank Zappa's favourite conductors and composers was Pierre Boulez, who's on the list of credits to the 1966 release, *Freak Out*. In 1983, Boulez commissioned Zappa to write three works for his small orchestra, the Ensemble Intercontemporain. At the same time, Zappa was composing on the Synclavier, which gave him the chance to add his computer-generated tracks to the resulting album [30].

The result was *The Perfect Stranger*, released in August 1984. Three months later another Synclavier-based album appeared, bearing the title *Francesco Zappa* and purporting to be a selection of pieces by a little-known 18th century composer of that name. It sounds like a joke—and many people took it that way—but it's the literal truth. There really was a classical composer named Francesco Zappa, he really was a contemporary of Haydn and Mozart, and the works on the album really are by him—arranged for Synclavier by his more famous 20th century namesake. Quoting John Corcelli again:

> The music was by an obscure Italian composer, Francesco Zappa, who lived from 1717 to 1803. He was a cellist and conductor who worked mostly in The Hague, in the Netherlands, and whose musical career supplemented his income as a teacher. Gail Zappa, Frank's wife, discovered a listing for the composer in *Grove's Dictionary of Music* ... Although they weren't related, Zappa was very interested and borrowed the Opus 1 trios and the Opus 4 sonatas by his namesake from the library. Next, Zappa put his Synclavier to work ... It took months of work to finish it, but the harmonic choices were limited, and the music isn't sophisticated enough to engage the listener. As a technical achievement, though, the album reinforced Zappa's confidence in *La Machine*, as he called it, as a means to an end [31].

It was mentioned a moment ago that Pierre Boulez was one of the many musicians credited as an influence on Zappa's first album, *Freak Out*. Another name on that list is Karlheinz Stockhausen—who, as we saw earlier in this chapter, was one of the key figures in the history of electronic music.

Frank Zappa wasn't the only rock musician to acknowledge Stockhausen's influence. We've come across a couple of other examples already, with Stockhausen being name-checked by Ray Manzarek of The Doors and Edgar Froese of Tangerine Dream. There's an even bigger name to add to list, too, as Philip Norman reveals in his biography of Paul McCartney. A friend of Paul's in the 1960s was Barry Miles, the proprietor of a trendy bookshop and avid follower of all the latest developments in classical music. As Norman explains:

> Miles thereafter introduced him to experimental music, whose cutting edge was not Anglo-American but European and whose leading figures made his and John's audio adventures seem tame indeed: the German Karlheinz Stockhausen, the Frenchmen Pierre Schaeffer and Edgard Varèse, and the Italian Luciano Berio. He learned how, following on from IBM's model 740 computer, machines

were turning into performers and magnetic tape into orchestras, cut up and reordered or played backwards or endlessly repeated on multi-layered loops [32].

That idea of tapes being "cut up and reordered or played backwards or end-lessly repeated on multi-layered loops" is something most people associate with the Beatles and other rock groups, but the fact is it was all done by avant-garde classical musicians first. Of course, it's an idea the Beatles picked up and ran with—although the band member who first latched onto it wasn't, as many people might guess, John Lennon. Quoting Philip Norman again:

> Whenever Paul saw John, he'd be full of news from this esoteric new world: how the tape-loop maestro Luciano Berio was coming to Britain to teach a course at Dartington Hall; how Paul had written a fan letter to Stockhausen—quite a turnaround for a Beatle—and received a personal reply. "God, man, I'm so jeal-ous," John would respond gloomily, but always left it at that [32].

But John caught up in the end. He was responsible for the Beatles' most Stockhausen-like work—the track "Revolution 9" from the White Album (1968), which musicologist Wilfrid Mellers describes as "an electronic freak-out and collage piece, distorting and mixing muzak of sundry kinds, with conversations and TV statements interspersed" [33]. Paul, on the other hand, remained dubious about the piece, as Norman explains in his biography:

> He argued that putting "Revolution 9" on the album was a piece of self-indulgence that would leave most listeners totally baffled [34].

Stockhausen's influence on rock music continued into later decades, crop-ping up, for example, in the work of the Icelandic singer Björk and the British band Portishead. For his own part, Stockhausen seems to have been flattered by the efforts of popular musicians to emulate him—but not terribly impressed by the results. When asked about Bjork and Portishead on a BBC documen-tary, he replied:

> They love the unusual sounds—the noisy sounds, crashing sounds, shooting sounds, they like that in my music, but they don't ask what they're made for or what the function of them is in my music. They just like the effect [35].

The fact is that, for Stockhausen and composers like him, the actual "sound" of a musical composition was only part of its significance. For him, the formal structure of a work—something that may not even be apparent to a casual listener—was equally important. Interestingly in the context of the present book, this formal structure, in the mid-20th century, was often influenced by explicitly "scientific" considerations—and that's the subject we'll turn to in the next chapter.

References

1. P. Griffiths, *Modern Music and After* (Oxford University Press, Oxford, 1995), p. 17
2. D. Leslie, G. Adamski, *Flying Saucers Have Landed* (Werner Laurie, London, 1953), pp. 104–105
3. D. Leslie, *Music of the Future* (CD liner and cover notes: Trunk Music, 2005)
4. C. Wen-Chung, *Varèse: The Complete Works* (CD liner notes, Decca, 2004)
5. P. Griffiths, *Modern Music and After* (Oxford University Press, Oxford, 1995), p. 45
6. R. Maconie, *Stockhausen on Music* (Marion Boyars, London, 1989), p. 90
7. R. Maconie, *Stockhausen on Music* (Marion Boyars, London, 1989), pp. 131–132
8. C. Deliège, *Conversations with Pierre Boulez* (Eulenberg Books, London, 1976), p. 109
9. D. Howe, M. Stammers, S. Walker, D. Who, *The Sixties* (Virgin, London, 1993), p. 12
10. P. Griffiths, *Modern Music and After* (Oxford University Press, Oxford, 1995), p. 207
11. Wikipedia article on "Musikalisches Würfelspiel", https://en.wikipedia.org/wiki/Musikalisches_W%C3%BCrfelspiel
12. W.A. Mozart, *Musikalisches Würfelspiel, K.516f.* Petrucci Music Libsrary. https://imslp.org/wiki/Musikalisches_W%C3%BCrfelspiel,_K.516f_(Mozart,_Wolfgang_Amadeus)
13. A. Lovelace, Analytical Engine. *Museum of Imaginary Musical Instruments.* http://imaginaryinstruments.org/lovelace-analytical-engine/
14. J. J. Coupling [John R. Pierce], Science for Art's Sake, *Astounding Science Fiction*, November 1950, pp. 83–92
15. J.R. Pierce, *The Science of Musical Sound* (Scientific American Books, New York, 1983), p. 211
16. J.R. Pierce, *The Science of Musical Sound* (Scientific American Books, New York, 1983), p. 11
17. J.R. Pierce, *The Science of Musical Sound* (Scientific American Books, New York, 1983), pp. 7–8
18. Wikipedia article on "Mellotron", https://en.wikipedia.org/wiki/Mellotron
19. P. Griffiths, *Modern Music and After* (Oxford University Press, Oxford, 1995), p. 68

20. M. Babbitt, *All Set* (CD liner notes, BMOP/sound, 2013)
21. R. Manzarek, *Light My Fire* (Berkley Boulevard Books, New York, 1999), p. 255
22. "An Avant-Garde Spirit for Tangerine Dream", *The New York Times*, October 1974 https://www.nytimes.com/1974/10/04/archives/an-avantgarde-spirit-for-tangerine-dream-the-pop-life.html
23. B. Eno, *Discreet Music* (liner notes, Virgin, 2009)
24. A. Walker, *Stanley Kubrick Directs* (Abacus, London, 1973), p. 292
25. S. Kubrick, *A Clockwork Orange* (Screenpress Books, Suffolk, 2000), pp. 311–312
26. P. Ramey, *A Clockwork Orange* (CD liner notes: Warner Bros, 2000)
27. D. Cronenberg, *Videodrome* (DVD back cover, Universal Studios, 2002)
28. Wikipedia article on "Videodrome (soundtrack)", https://en.wikipedia.org/wiki/Videodrome_(soundtrack)
29. J. Corcelli, *Frank Zappa FAQ* (Backbeat Books, Milwaukee, 2016), p. 143
30. J. Corcelli, *Frank Zappa FAQ* (Backbeat Books, Milwaukee, 2016), pp. 123–124
31. J. Corcelli, *Frank Zappa FAQ* (Backbeat Books, Milwaukee, 2016), p. 145
32. P. Norman, P. McCartney, *The Biography* (Weidenfeld & Nicholson, London, 2016), pp. 225–226
33. W. Mellers, *Twilight of the Gods: The Beatles in Retrospect* (Faber, London, 1976), p. 135
34. P. Norman, P. McCartney, *The Biography* (Weidenfeld & Nicholson, London, 2016), p. 3
35. J. Needham, The Cosmic Messenger: How Karlheinz Stockhausen shaped contemporary electronic music, *The Vinyl Factory*, November 2017. https://thevinylfactory.com/features/karlheinz-stockhausen-electronic-music-influence/

Scientific Music

From the middle of the 20th century onwards, musicians increasingly looked to the world of science for new ideas. In this chapter we look at the influence of science on new musical trends such as the serialism of Stockhausen and Boulez and the "stochastic music" of Xenakis. Scientists like Einstein and Heisenberg were cited as sources of inspiration in the way that poets or philosophers might have been in an earlier age. By the end of the 20th century, musical works were being composed using real astronomical data—as in Fiorella Terenzi's *Music from the Galaxies*—or, moving from outer space to the Earth's oceans, the songs of humpback whales.

A New Perspective

For hundreds of years—all the way back to the time of Pythagoras, in fact—musicians had been happily getting on with their work without needing to delve too deeply into the scientific and mathematical side of their subject. In the 20th century, however, the situation changed abruptly with the advent of electronic music—which, as we saw in the previous chapter, required a working knowledge of things like sine-wave oscillators and Fourier synthesis in its early days.

This enforced encounter with science changed the very way modernist composers thought about music in the 1950s. They started to talk about the "parameters" of music, meaning all the things that a composer has control

© The Editor(s) (if applicable) and The Author(s),
under exclusive licence to Springer Nature Switzerland AG 2020
A. May, *The Science of Sci-Fi Music*, Science and Fiction,
https://doi.org/10.1007/978-3-030-47833-9_4

over—from the pitch, duration and volume of notes to the speed, or tempo, at which music is played and the instruments it's played on.

Up to this point, the most serious attempt to "mathematize" music had been Schoenberg's 12-tone system, as discussed in the second chapter. While Schoenberg focused on just a single parameter, pitch, it occurred to a number of composers of the following generation that a similar formalism could be applied to all the other parameters as well. The resulting experiments became known as "serialism"—and for the most part they're more memorable for the cleverness of their structure than for their aesthetic appeal.

One of first serial works, dating from 1951, was *Structures* by Pierre Boulez. He's been mentioned already as one of the pioneers of electronic music, but this particular work was written for an ordinary piano. What's extraordinary about it, as the title suggests, is its construction. A famous analysis of the piece was produced a few years later by someone we met in the first chapter— György Ligeti, whose music was used to such good effect in *2001: A Space Odyssey*. Regarding Ligeti's dissection of *Structures*, musicologist Paul Griffiths calls it "as much a classic document as the piece itself". He goes on:

> As Ligeti demonstrates, Boulez obtained his duration series by applying numbers to the pitch series, and then translating all the other serial forms into number sequences by using the same pitch-number equivalences. [1]

Another composer who experimented along these lines was Stockhausen, one of the major protagonists of the previous chapter. In his case, he was drawn to the serialist approach for scientifically argued reasons as much as aesthetic ones. To quote Griffiths:

> Work on his electronic *Studien* had given Stockhausen a practical demonstration of how pitch and duration are aspects of a single phenomenon, that of vibration. A vibration of, say 32 Hz will be perceived as a pitched note, whereas one of 4 Hz will be heard as a regular rhythm, and somewhere in between the one will merge into another. So … some deep coherence had to be sought between the principles applied to pitch and to rhythm in forming a work. [2]

As impeccable as Stockhausen's reasoning looks, a sceptic might say he's swung so far over to the scientific side that he's no longer talking about music. It's true that, from a purely scientific perspective, the only difference between a "rhythm" and a "note" is where it happens to lie on the frequency scale. But the fact remains that our ears perceive them as completely different musical effects. It's far from obvious, therefore, that the same organizing principle should be applied to both.

Still, it's an interesting experiment—and Stockhausen was such an intuitively brilliant musician that virtually everything he produced is worth listening to, whether it proved the point he was trying to make or not. One of his best known works from the 1950s is *Gruppen*, which—despite its electronic inspiration—is written for a traditional orchestra, albeit one that's divided into three sections that play almost independent of each other.

If you search online you can find numerous scientific-looking diagrams showing how *Gruppen* is structured around "groups" of different musical parameters—a subject there's fortunately no need to go into detail on here. The brief snippet shown in Table 1 is enough to give a flavour of what's involved.

It's significant that the composers who championed serialism in the 1950s were, for the most part, the same people who were exploring the possibilities of electronic music at the same time. Europe had Boulez and Stockhausen, while America had Milton Babbitt—another pioneer of electronic music who was mentioned in the previous chapter. Babbitt advocated, in his own words, "applying the pitch operations of the 12-tone system to non-pitch elements: duration, rhythm, dynamics, phrase rhythm, timbre and register, in such a manner as to preserve the most significant properties associated with these operations in the pitch domain when they are applied in these other domains" [4].

It may sound as though Babbitt is simply echoing his European contemporaries here, but there's a subtle difference in approach. Instead of using serialist logic to create an ad-hoc framework from the bottom up for each new composition, as Stockhausen and Boulez did, Babbitt preferred a top-down approach—using serialism, as he put it, "as a means of characterizing or discovering general systematic, pre-compositional relationships" [5].

That sounds admirably scientific, but many readers will be sceptical that the underlying mathematical relationships Babbitt is talking about would ever

Table 1 The musical "parameters" for one section of Stockhausen's *Gruppen* [3]

Group	23	24	25	26	27	28	29	30	31	32
Pitch-class	F (5)	D# (3)	E (4)	G (7)	D (2)	F# (6)	C (0)	G# (8)	C# (1)	A# (10)
Octave	3	4	5	3	3	4	5	3	4	4
Note length	Half	Quarter	Eighth	Half	Half	Quarter	Eighth	Half	Half	Eighth
Tempo (BPM)	107	95	101	120	90	113.5	80	63.5	85	71

be apparent to the listener. In fact is they hardly ever are—but they were never meant to be, because the mathematics was only a means to an end. The fact that listeners are unaware of the mathematical substructure, however, doesn't mean there was no need for it in the first place. As Paul Griffiths says:

> The combinatorial relationships and the rhythmic serialism may not be noticed as such when the music is played and heard, but they surely contribute to the impression of lucidity, elegance and pleasure the piece conveys. There is something watertight about the music; every detail has an immediate answer, suggesting that every eventuality has been prepared by the composer. [6]

The reality is that great music is something that comes out of a great composer's head—and many things, including mathematics, may serve as a catalyst to that. It's a point that was made in, of all places, the November 1950 issue of *Astounding Science Fiction* magazine. Another person we met in the previous chapter, the engineer and computer music pioneer John R. Pierce, wrote the following in a non-fiction article, printed under the pen-name of J. J. Coupling, entitled "Science for Art's Sake":

> A sceptic might argue that a composer of genius can make a good thing of anything. Certainly, when mathematics is used merely as a sort of guide or crutch, it is hard to apportion credit between the mathematics and the user. [7]

While the word "mathematics" may suggest extreme precision, another way it came to be used was just the opposite, through the use of random numbers and statistical techniques. This subject was touched on briefly in the first chapter, when we saw how John Cage used coin tosses to string together electronic fragments to create *Williams Mix*. On other occasions, for example in the piano piece *Music of Changes* (1952), Cage used the ancient Chinese oracle *I Ching*—familiar to SF readers from Philip K. Dick's novel *The Man in the High Castle* (1962)—instead of coin tosses.

An even stranger example of a random—or at least pseudo-random—musical composition is Cage's orchestral work *Atlas Eclipticalis* (1962). It shares its title with an astronomical atlas published in 1958, which Cage used to generate notes for the piece in a way that was totally characteristic of him. As astronomer Andrew Fraknoi explains, "He put see-through musical notation paper in front of a star atlas, and let the positions of the brighter stars in the atlas determine the positions of the notes on the paper" [8].

John Cage was just one of many composers to experiment with random musical composition, which became common enough in the 1950s to be given a name: "aleatoric music", from the Latin word *alea* meaning dice. In the context of the tune-based music of the 18th and 19th centuries, the idea of randomness would have been bizarre, but with the emergence of predominantly texture-based music in the 20th century it makes much more sense. Here is Stockhausen on the subject:

> In classical music, the most important principle was that you should always hear everything—every individual tone—so that one wrong note in a chord was immediately noticeable. In statistical compositions, however, the individual components enter into textures that have their own overall characteristics and become new units which are treated like sounds… You can exchange the position of elements within given limits at random and it doesn't change the characteristics. Like changing the positions of a tree's leaves. You can say "this is a beech tree," even if all the leaves have changed their position. [9]

For Stockhausen, the use of randomness in music was rather more sophisticated than Cage's coin tossing, as can be seen from the following description of his early experiments with electronic music:

> I would give my three collaborators each a sheet of paper with a curve drawn on it, all of them to be executed in 20 seconds, to make a certain sound-event. And I would say to the first collaborator, this time start the pulse generator at 4 pulses per second, follow the curves of the drawing and end up with 16 pulses per second. To the second, who was working the potentiometer controlling the loudness level, I would say, let us take the dynamic range as being 4 decibels, from this maximum to this minimum, following the curve. To the third assistant who was in charge of the electronic filter, which lets through only a narrow band of frequencies from the signal, I would say, start at 3,000 Hz, follow this curve for 20 seconds, and finish up at 400 Hz. [10]

A scientist might complain that this is nothing but pseudoscience, using scientific-sounding terminology in a way that has no intrinsic physical meaning. That's true enough, but it makes no difference to the outcome. For the composer, mathematics—as we saw a moment ago—is just a means to an end. There's no reason it has to be "scientific" mathematics—it could equally well be pseudoscientific or mystical numerology. The end result, depending on the talent and imagination of the composer, can still be great music.

The archetypal example of "non-scientific" mathematics in music is the use of magic squares by the former Master of the Queen's Music, Sir Peter Maxwell Davies (1934–2016). A magic square is an $n \times n$ array of numbers, all different, which add up to the same total in each row, column and diagonal. Arrays of this type have no scientific significance—or musical significance, for that matter—but Davies used them as a basis for numerous compositions, applying the array's numbers to pitch classes and note durations in way closely analogous to serialism.

Works composed in this way include his Symphony no. 1 from 1976 [11] and the *Sinfonietta Accademica* from 1983 [12], as well as a short but spooky opera, *The Lighthouse*, which had its premiere in 1980 [13]. The latter is based on a real-world incident, the unexplained disappearance of three men from the Flannan Isles lighthouse in 1900, which also provided the inspiration for the *Dr Who* serial "Horror of Fang Rock" (1977) and the sci-fi video game *Dark Fall: Lights Out* (2004). In Davies's opera the men are convinced they are haunted by ghosts, but whether these are real or imaginary is for the audience to decide. As Davies himself wrote:

> There is the further possibility that we have been watching a play of ghosts in a lighthouse abandoned and boarded up for 80 years. [14]

Another of Davies's magic-square pieces with a supernatural connection is *A Mirror of Whitening Light* (1977), the title of which refers to the supposed alchemical process by which base metals are transformed into gold. Here is a description of it from a musicological point of view by Paul Griffiths:

> Suitably enough the agent of the work, in the alchemical sense, is the magic square of Mercury (in mathematics a square formed of 8 × 8 digits, the sums of whose rows of numbers are the same, read horizontally and perpendicularly). This, suitably used, can generate perfectly recognizable and workable sequences of pitches and rhythmic lengths, easily memorable once the key to the square has been found. [15]

Despite the significance of the element mercury in alchemy, the magic square of Mercury (Fig. 1) actually refers to the planet—one of seven such squares traditionally assigned to heavenly bodies.

The use of the magic square in *A Mirror of Whitening Light* is explained further in online class notes by Gareth E. Roberts:

> Davies takes 8 notes from the plainchant *Veni Sancte Spiritus*, often called the "Golden Sequence", to create a melodic phrase with 8 distinct notes. He then transposes the phrase to start on each of the 8 notes, yielding an 8 × 8 matrix of notes. This matrix is then mapped onto the magic square of Mercury. Different paths through the resulting magical music square generate the pitches and rhythmic durations for the piece. [16]

8	58	59	5	4	62	63	1
49	15	14	52	53	11	10	56
41	23	22	44	45	19	18	48
32	34	35	29	28	38	39	25
40	26	27	37	36	30	31	33
17	47	46	20	21	43	42	24
9	55	54	12	13	51	50	16
64	2	3	61	60	6	7	57

Fig. 1 The magic square of Mercury, as used by Peter Maxwell Davies in *A Mirror of Whitening Light*

As Roberts says, "analysing the patterns in *A Mirror of Whitening Light* is a bit like doing a musical word-search". For example:

The durations of the notes of the flute part form the sequence 7, 6, 4, 5, 3, 2, 8. These are the numbers in the top row of the magic square [backwards] reduced modulo 8, with zero replaced by an 8. The rhythmic durations for the flute part follow the same spiral pattern through the magic square as the notes! [16]

Coupled with the increasing use of electronics, this emphasis on the numerical aspects of music might have left non-technically minded composers feeling left behind. With this in mind, in 1977 Pierre Boulez created the Institute for Research and Coordination in Acoustics and Music—IRCAM for short—at the Pompidou Centre in Paris. Its aim was to bridge the gap between science and music—"a unique location where artistic sensibilities collide with scientific and technological innovation", according to its website [17].

As well as providing facilities for composers who are already tech-savvy, IRCAM has resident staff members to assist those without a technology background to realize whatever electronic or computer-assisted music they may have in their heads. From a science-fictional point of view, one of the most important works to come out of IRCAM was Tod Machover's opera *VALIS*, based on Philip K. Dick's novel of the same name. The opera was commissioned for the tenth anniversary of IRCAM in 1987—and will be discussed in more detail in the last chapter of this book, "Speculations on a Musical Theme".

Although its main aim is to support musicians with no programming knowledge, IRCAM can also benefit software geeks who lack musical training, via its GNU-licensed software package OpenMusic [18]. Free to download and use, this allows anyone with a basic understanding of computer programming to compose quite sophisticated-sounding music (although whether the result has any aesthetic value is another matter). Going back to the idea of serialism, a simple example of its use is shown in Fig. 2, which generates a random 12-note sequence in which both pitches and note lengths are derived from the same basic series.

Over the years, the OpenMusic community has supplemented it with an impressive range of library functions. A very basic one—*retrograde*, which simply reverses the order of notes—appears in the foregoing example, but others can be much more sophisticated. There's even a library function called *luxaeterna*, which allows the user to create a random piece of music that sounds a little like Ligeti's *Lux Aeterna*, as featured in *2001: A Space Odyssey*.

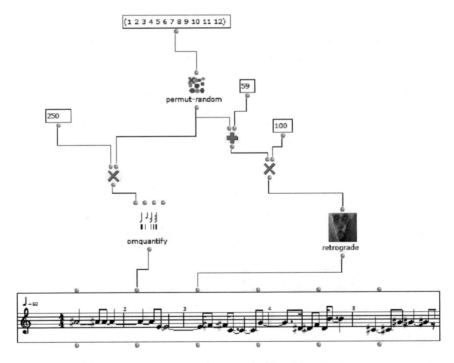

Fig. 2 A simple example of "serial composition", using OpenMusic to create a random tune with note durations based on a 12-element series, and pitches based on the same series backwards

Science-Inspired Music

While the language and tools of contemporary music can look superficially scientific, there's still a huge difference between art and science in that the latter strives to be objective, while the former is ultimately always subjective. This places it closer to pseudoscience, whose practitioners can choose "theories" to suit their personal tastes. It's an analogy that Pierre Boulez made, when he was asked if composers had become research scientists:

> I should think that research scientists are a lot less speculative… That is, they are less likely to be hypnotized by a construction that may satisfy them from the intellectual point of view. The real research workers are precisely those who are constantly in touch with reality, and who modify their working hypotheses according to their findings. The pseudoscientist, on the other hand, will invent a construction for his own satisfaction without being too concerned about relating it to reality. [19]

The fundamentally subjective nature of music, as compared to science, was also stressed by another composer, Charles Wuorinen:

> In science, observations of phenomena are made and generalized inferences are drawn from the observations. A theory then emerges, which has the power to predict future events… In science, the theory can be tested by external criteria; if it doesn't predict correctly, it is discarded. In art also, theory usually arises from observation—but … theory here is a generalization from existing art works, which, however, are neither validated nor invalidated by the theory. [20]

Be that as it may, people still love to draw parallels between science and art—and Boulez himself was no exception. With reference to Schoenberg's 12-tone method of composition, he wrote that "with the 12-tone system, music moved out of the world of Newton and into the world of Einstein" [21].

Einstein, of course, was the most famous scientist of the 20th century, so it's no surprise that Boulez—and many others—should pick him out for comparison with Schoenberg. As Brian Stableford writes in his book *Science Fact and Science Fiction*:

> The mathematical foundations of music were shaken in the early 20th century by the atonal music of Arnold Schoenberg, whose Chamber Symphony no. 1 was composed in 1905, the year that Albert Einstein published the special theory of relativity—a coincidence that seemed sufficiently significant to some observers for both men to be asked to comment on the apparent affinity. [22]

It's true that Einstein and Schoenberg did have several things in common. They were born within five years of each other to Jewish parents living in German-speaking countries—the former in Germany, the latter in Austria—and they both emigrated to America in the 1930s when Europe became such an uncomfortable place for Jews. They shared concerns about the future of the Jewish nation, and corresponded with each other on the subject in the 1920s [23]—as well as meeting face-to-face at a Jewish fundraising event in New York in 1934 [24].

Despite all the speculation, however, there's no real evidence that Schoenberg's music was directly influenced by Einstein's work. It's something he denied, in fact, as the following news report on his arrival in America in 1933 shows:

> Surrounded by the gentlemen of the press and photographers last week, Arnold Schoenberg, the newest musical luminary come to our shores, endeavoured to answer the questions hurled at him. These ranged from the food he prefers to a problematic relationship between the science of mathematics as expressed by Einstein and the science of music as developed by Schoenberg. This last question particularly amused the eminent composer. "There may be a relationship in the two fields of endeavour," he said, "but I have no idea what it is. I write music as music without any reference other than to express my feelings in tone. I do not shape my scores with a definite idea to express anything but music." [25]

For his part, Einstein's tastes in music were surprisingly old-fashioned—his favourite composers were Bach, Mozart and Schubert—and he was utterly baffled by Schoenberg's music, which he described as "crazy" [24]. In 1976, just over 20 years after his death, he became the subject of a modernist opera himself: Philip Glass's *Einstein on the Beach* (Fig. 3). Here's what the BBC's music website has to say about it:

> It is well known that Einstein was fond of music, playing the violin and piano while enjoying Bach and Beethoven's work. But as someone who was left cold by Debussy and Wagner, one can only imagine what he might have thought of Glass's first opera, *Einstein on the Beach*. The work gets rid of usual orchestral arrangements of operas in favour of simply synthesizers, woodwind and voices, and has four acts stretching over five hours, without a specific plot, but making reference to events throughout Einstein's life. [26]

One composer who was never shy about drawing parallels between science and music was Stockhausen, who did so time and again throughout his career. Often they're parallels which, from a scientist's point of view, seem tenuous to

Fig. 3 A performance of Philip Glass's opera *Einstein on the Beach* in Dortmund in 2017 (Wikimedia Commons user Sky Xe, CC-BY-SA-3.0)

say the least. Consider the following example, where Stockhausen attempts to link Schoenberg's third String Quartet from 1927 with Einstein's quest for a unified field theory:

> There was a general feeling at this time in history that scientists were very close to announcing the formula that was underlying the entire universe, the well-known unifying formula we often hear in connection with the name of Einstein. This simply says that everything is a development of one thing, which is exactly what Schoenberg says in that one composition. [27]

As an aside, it's worth pointing out that Schoenberg's own account of the genesis of that particular string quartet sounds closer to a horror movie than to fundamental physics:

> As a little boy I was tormented by a picture of a fairy-tale, "The Ghost Ship", whose captain had been nailed through the head to the topmast by his rebellious crew... It might have been, subconsciously, a very gruesome premonition which caused me to write this work, because as often as I thought about this movement, that picture came to mind. [28]

Another of Stockhausen's over-the-top assertions relates to one of his own compositions, *Mantra* (1970):

As it stands, *Mantra* is a miniature of the way a galaxy is composed. When I was composing this work … it demanded itself, it just started blossoming. As it was being constructed through me, I somehow felt that it must be a very true picture of the way the cosmos is constructed. [29]

It's not at all clear what Stockhausen meant by "a miniature of the way a galaxy is composed"—but even so, *Mantra* does have significant interest from a scientific point of view. It's scored for two ordinary acoustic pianos, but their sound is passed through a ring modulator and projected to the audience through loudspeakers. The resulting effect—which is undoubtedly "spacey", whether or not it's an accurate model of the galaxy—is described in the liner notes of the Naxos CD as follows:

Central to *Mantra* is the electronic transformation of the piano sounds using ring modulation, whereby the sound of each piano is modulated with a synthetic tone tuned to the central pitch of each section. This is the slowest mantra of all, whose pitches are heard not directly, but only through the transformation of the piano timbres. The degree of sonic transformation depends on the interval between the synthetic tone and the piano note. With simple intervals (octaves, fifths) the transformation is subtle, but with more complex intervals (semitones, major sevenths) it is more dramatic, resulting in metallic, bell-like sounds and quasi-vibrato. [30]

Stockhausen was also fond of name-dropping scientists in the context of his own musical experiments, referring for example to "the strong influence on musicians, during the early fifties, of certain books for the general reader by Einstein or Heisenberg" [31]. That's not quite as glib as it might sound, because Stockhausen wasn't talking about the specific physical context of the science, but the general principle by which, as he said on another occasion, "statistical processes became very important":

The Heisenberg uncertainty principle is based on this hypothetical behaviour of the components of the atom. That was the main thing in the air at the end of the 40s and beginning of the 50s. We worked with theories in communication science; Shannon was very important as a mathematician—Markov too. And I simply transposed everything I learned into the field of music and for the first time composed sounds which have statistical characteristics. [32]

A contemporary of Stockhausen who also took inspiration from statistical theory was the Greek composer Iannis Xenakis (1922–2001). He was mentioned briefly in the previous chapter as one of the first people to use computer programs to assist with the writing of music. Xenakis coined the term "stochastic music" to describe his approach to composition—stochastic being a term often used in science and engineering to describe random processes. Significantly, Xenakis originally trained as engineer himself. As music critic Paul Driver wrote in 1994:

> One of his chief musical contributions has been the application to composition of probability theory, by means of computer programming. Unlike John Cage, who opened the floodgates of randomness and pure chance on music, Xenakis exploits mathematical, or what he properly calls "stochastic", principles in order to harness chance to carefully planned purposes. [33]

Xenakis originally adopted this approach as a reaction against the highly deterministic serialist method of Boulez and Stockhausen, which struck him as an unnecessarily complicated way to create a sound that—as far as listeners were concerned—was simply "a mass of notes in various registers", as he put it. Paul Griffiths explains his reasoning:

> For him, the conclusion had to be that another theoretical base was needed. If the effect was to be nothing but a mass of notes, the means to produce that effect should be sought in the branch of mathematics that had been developed to deal with such statistical phenomena: in "the notion of probability, which implies in this particular case, combinatory calculus". He therefore turned to the laws of stochastics, which describe phenomena that can be defined only in the large—and so derived "stochastic music". [34]

In 1963, Xenakis produced a book called *Formalized Music: Thought and Mathematics in Composition*—which is still considered influential enough to have its Wikipedia entry [35]. As that article states, "the book contains the complete FORTRAN code for one of Xenakis's early computer music composition programs". Yet the book remains the work of Xenakis the musician, not Xenakis the engineer. As John Pierce puts it in his own book, *The Science of Musical Sound*:

> The mathematics in his *Formalized Music* seems disjointed and sometimes irrelevant to this engineer, and would throw any non-mathematical musician off, but it is apparently of use to Xenakis, for he composes very attractive music. [36]

In last two decades of the 20th century, statistical mathematics encroached even further into the world of the arts with the work of people like Feigenbaum and Mandelbrot on "fractals"—visually attractive patterns that replicate themselves when zoomed in to higher magnification—and the wider topic of "chaos theory". The latter is something of a misnomer, because the processes it deals with are "chaotic" only in the sense that they are unpredictable. The results, however, are often intricately structured in a way that's as appealing to artists as they are interesting to scientists. One musician who was particularly impressed by such developments was György Ligeti, who wrote:

> Further influences that enriched me come from the field of geometry—pattern deformation from topology and self-similar forms from fractal geometry—whereby I am indebted to Benoit Mandelbrot. [37]

The OpenMusic software mentioned earlier comes with a library called *OMChaos*, which contains modules that can generate strings of numbers according to standard chaos-theory algorithms such as the Henon and Lorenz attractors. Another built-in library, *OMAlea*, provides various "aleatoric" functions which are more likely to be familiar to engineers and scientists than musicians, such as Poisson, Gaussian and Weibull distributions. It also includes several modules relating to Markov chains, which are particularly useful in composing realistic-sounding "random" music.

Eagle-eyed readers will already have spotted the name Markov—referring to the Russian mathematician Andrei Markov (1856–1922)—in one of the Stockhausen quotes earlier. His great discovery was that, in many statistical processes, the probability that a particular state will arise depends only on the immediately preceding state. He was thinking about natural process, where this can sometimes be a very good approximation.

On the other hand, when applied to music—which is a long way from its original context—it's certainly not true that, say, the pitch of a note depends only on the pitch of the previous note. Nevertheless, when creating a computer composition, a Markov model can produce a much more realistic result than pure randomness.

By analysing the note sequence in an existing composition, or a whole body of compositions that one wants to emulate, it's possible to construct a transition matrix (such as the example illustrated in Fig. 4) showing the probability with which each note is followed by any other note. Selecting notes according to these probabilities will then produce a more convincing-sounding tune than if the notes had simply been picked at random.

	C#	D	D#	E	F	F#	G	G#	A	A#	B	C
C#	0	0	0	1	0	0	0	0	0	0	0	0
D	0	0	0	0	0	0	1	0	0	0	0	0
D#	0	0.333	0.333	0.333	0	0	0	0	0	0	0	0
E	0	0	0.118	0.471	0	0	0.118	0	0.059	0.059	0.118	0.059
F	0	0	0	0.5	0	0.5	0	0	0	0	0	0
F#	0	0	0	0.2	0.4	0.4	0	0	0	0	0	0
G	0	0.077	0	0	0	0.154	0.692	0	0.077	0	0	0
G#	0	0	0	0.667	0	0	0	0	0.333	0	0	0
A	0.125	0	0	0.125	0	0	0	0.375	0.375	0	0	0
A#	0	0	0	0.333	0	0	0	0	0	0	0.667	0
B	0	0	0	0.167	0	0	0	0	0.167	0.333	0.333	0
C	0	0	0	0	0	0	0	0	1	0	0	0

Fig. 4 Example of a Markov transition matrix—in this case, based on the Pink Floyd song "Interstellar Overdrive". The idea is that, after an initial note is chosen at random, the user finds that note among the red notes on the left, then selects the next note from the green notes along the top based on the probabilities shown in the matrix. This new note then becomes the "red note" for the next iteration, and the process is continued as long as required

Music from Outer Space

As we saw in the second chapter, "Musical Mathematics", the idea of an ethereal "music of the spheres" permeating the cosmos has been around, in one form or another, since ancient Greek times. One of its most famous proponents was the astronomer Johannes Kepler, who wrote on the subject at great length in his book *Harmonices Mundi* (1619). That title translates as "The Harmony of the World"—or in German *Die Harmonie der Welt*, which also happens to be the title of a 1957 opera about Kepler's life by the composer Paul Hindemith.

Probably of more appeal to contemporary audiences, however, is another opera on the same subject—this one the work of Philip Glass, and simply titled *Kepler*. Dating from 2009, Glass had moved on by this time from the uncompromising modernism of *Einstein on the Beach*, and the music of *Kepler*—a mere two hours in length, compared to *Einstein*'s five—is much more accessible. It's almost "easy listening", in fact, in a similar style to Glass's *Low* and *Heroes* symphonies from the 1990s, which were inspired by David Bowie's albums of those titles.

Actually *Kepler* is scarcely an opera in the conventional sense, since there's no action to speak of—just a long philosophical argument between the forward-looking Kepler and his more conservative contemporaries. The only reference to the "music of spheres" takes the form of a quotation from Kepler's writings:

> The celestial motions are nothing else than a continuous heavenly music which can be perceived only with the mind, not with the ear. [38]

Put this way, Kepler's notion of celestial harmony is nothing like as outlandish as it's sometimes made out to be. All he's saying is that numerical relationships exist between the motions of planets which are analogous to those found in musical harmony. He's not claiming these relationships produce music in the literal sense, because they're "perceived only with the mind, not with the ear".

There's another way in which the analogy is approximate, rather than exact, because planets go round and round in orbit, rather than vibrating back and forth like the strings of musical instruments. In both cases, however, it's meaningful to talk about frequencies—and ratios between frequencies—and this seems to be what Kepler was getting at.

For a planet moving in a perfectly circular orbit, the frequency is easy to calculate: it's simply the reciprocal of the time taken to complete one revolution. But one of Kepler's great discoveries was that planets actually follow elliptical paths, at speeds that vary around the orbit (being faster when the planet is closer to the Sun). Kepler imagined that the frequency—or "pitch", in the musical sense—of a planet was simply its instantaneous angular speed, measured in degrees per second, divided by 360.

This gives the frequency in Hertz (although Kepler, who lived three centuries before Heinrich Hertz, wouldn't have put it that way)—and it's a very low frequency indeed. The Earth, for example, takes a year to complete an orbit—which corresponds to a frequency of just three hundredths of a millionth of a Hertz. You couldn't hear that even if it was real music.

To translate Kepler's planetary harmonies into something that's audible, the whole system has to be scaled up in frequency several million times. That's exactly what scientist John Rodgers and musician Willie Ruff did, as an intellectual exercise, back in 1979. They chose an arbitrary scaling factor such that Earth's real-world frequency of 1 revolution per year becomes 800 Hertz in the scaled version. There's a lot of musical jargon in their description of the result, but it's worth quoting a sizeable chunk of it just to get the general gist:

> Mercury, as the innermost planet, is the fastest and the highest pitched. It has a very eccentric orbit, which it traverses in 88 days; its song is therefore a fast whistle, going from the E above the piano down more than an octave to about

C-sharp and back, in a little over a second. Venus and Earth, in contrast, have nearly circular orbits. Venus's range is only about a quarter-tone, near the E next above the treble staff; Earth's is about a half-tone, from G to G-sharp at the top of that staff… Next out from Earth is Mars, again with an eccentric orbit… It ranges from the C above middle C down to about F# and back, in nearly 10 seconds. The distance from Mars to Jupiter is much greater than that between the inner planets, and Jupiter's song is much deeper, in the baritone or bass, and much slower. It covers a minor third, from D to B just below the bass staff. Still farther out and still lower is Saturn, only a little more than a deep growl, in which a good ear can sometimes hear the individual vibrations. Its range is a major third, from B to G, the B at the top being just an octave below the B at the bottom of Jupiter's range. Thus the two planets together define a major triad, and it may well have been this concord … that made Kepler certain he had cracked the code and discovered the secret of the celestial harmony. [39]

The same logic by which planetary motions can be scaled up in into audible sounds applies equally well in the opposite direction. Most of what we know about the universe comes from electromagnetic waves, and these can be scaled down in frequency until they become audible. As well as being a novel way to create "space music", converting signals from, say, a radio telescope into sounds can provide useful scientific insights as well.

The science of "acoustic astronomy" was created—more or less single-handedly—in 1987 by Fiorella Terenzi (Fig. 5), as part of her doctoral research at the University of California's Computer Audio Research Laboratory. On her website, she explains how she began with radio observations of the distant galaxy UGC 6697:

To convert the radiation's frequency and intensity to audible form, I needed a special computer sound synthesis program which I used to elaborate the signal in terms of sound. After a variety of processing, this signal can be sent to a digital-to-analog converter and played through conventional loudspeakers, or recorded onto digital tape or CD, to bring you the sound of UGC 6697 from 180 million light years away. [40]

Terenzi is now a professor of physics and astronomy at Florida International University, and her web page at that institution includes audio samples—suitably rescaled—of numerous astronomical phenomena from the Sun, the Earth's magnetosphere and the planets Jupiter and Saturn to the Vela and Crab pulsars, the X-Ray binary GRS 1915+105—and of course galaxy UGC 6697 [41].

Fig. 5 Astrophysicist and science popularizer Fiorella Terenzi developed the technique of "acoustic astronomy" while working on her doctoral thesis, and has since produced numerous musical works derived from astronomical data (public domain image)

It goes further than that, however. In addition to her academic work, Terenzi has a sideline in science outreach—to "entertain, educate, enlighten and enthral" [42], in her words—as well as formal musical training. So it's only natural that she combined all these different interests to create a full-blown musical work from her recording of UGC 6697, which Island Records released as *Music from the Galaxies* in 1991. Since then her musical career has continued in parallel with her scientific one, as her biography on the Florida International University website explains:

> As a recording artist, she has recorded "Quantum Mechanic" and "NEO", tracks for Billboard Top 20 Music Video *The Gate to the Mind's Eye* with Thomas Dolby (Giant/Warner Bros Records); *Beyond Life*, a Mercury Records dance/trance CD tribute to Dr Timothy Leary (Polygram Records); *Trance Planet Vol. 5* (Triloka Records) and others. [42]

Although the main focus of this book is on the 20th century, it's interesting to take a brief look at a few more recent excursions into "acoustic astronomy". In 2011, for example, astronomers Alex Parker and Melissa Graham produced a *Supernova Sonata*, the musical parameters of which are taken from a series of Type 1a supernova events—241 in all—observed by the Canada-France-Hawaii Telescope between April 2003 and August 2006. As described on Parker's website (which also contains an embedded recording of the piece):

Each supernova is assigned a note to be played:
- Volume = distance. The volume of the note is determined by the distance to the supernova, with more distant supernovae being quieter and fainter.
- Pitch = stretch. The pitch of the note was determined by the supernova's "stretch", a property of how the supernova brightens and fades. Higher stretch values played higher notes. The pitches were drawn from a Phrygian dominant scale.
- Instrument = mass of host galaxy. The instrument the note was played on was determined by the properties of the galaxy which hosted each supernova. Supernovae hosted by massive galaxies are played with a stand-up bass, while supernovae hosted by less massive galaxies are played with a grand piano. [43]

A similar exercise was reported on a NASA blog—again with linked online recording—in 2012 [44]. The data this time came from a gamma ray burst, which was converted into music by Sylvia Zhu. Each gamma-ray photon is scaled all the way down from its original, enormously high, frequency to the audible spectrum, and the rate of arrival of photons was also slowed down to a sensible musical tempo.

Even more recently, astronomers C. A. Droppelmann and R. E. Mennickent produced a paper entitled "Creating Music Based on Quantitative Data from Variable Stars" in November 2018 [45]. The specific star they chose for their example composition (which again can be heard in an online recording) was RV Tauri, which fluctuates in magnitude in a way that is neither completely regular nor completely random.

The authors devised a mathematical formula to translate the star's varying magnitude into a musical pitch, dividing the range between minimum and maximum magnitude into 24 equal steps, and then assigning each of these to the 24 semitones in a two-octave range. To make the resulting composition more interesting, it's played as a four-part "canon"—a traditional form of music in which four groups of instruments play the same melody, but in different octaves and with offset entries:

The melodic line, the music derived from the star light curve, is introduced by the first violins in the third measure. In the fourth measure, the second violins begin the canon as the second voice. Then the violas begin in the fifth measure and finally, the cellos in the sixth measure. From there, the musical body is developed by following the melodic line derived from the star, until the last musical note corresponds to the last magnitude measured. [45]

Until intelligent life is discovered elsewhere in the universe, we can only speculate on what genuinely extraterrestrial music might sound like. Nevertheless, there's a good contender for "alien music" right here on Earth, in the form of whale songs (see Fig. 6). As far as we can tell, these are the product of creativity and intelligence—but of a non-human kind. Like the experiments in acoustic astronomy just mentioned, it's a topic that combines scientific interest with aesthetic appeal.

Although they've presumably been around as long as the whales themselves, whale vocalizations only really impinged on scientists' awareness during the Cold War, when a lot of effort went into listening to underwater sounds as part of the hunt for the other side's submarines. Then in 1970 the subject crossed over from the scientific into the artistic domain, when bio-acoustician

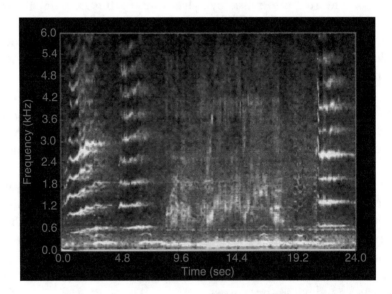

Fig. 6 The spectrogram of a humpback whale song, showing acoustic intensity as a function of frequency—running from 0 to 6 kHz on the vertical axis—and time, running from 0 to 24 seconds on the horizontal axis (Wikimedia Commons user Spyrogumas, CC-BY-SA-3.0)

Roger Payne released a long-playing record called *Songs of the Humpback Whale*, consisting solely of whale-produced sounds.

Payne's album couldn't have been better timed, coinciding as it did with a peak in the popularity of alien, spacey-sounding music—or as a 2014 piece in *Wire* magazine put it, "the very moment the world was most open to sounds from the unknown: the psychedelic and the trippy". The article goes on:

> Humpback whale song fit the bill perfectly. From high wails to deep growls to rhythmic scratches to tearful moans, it encompasses the full range of emotions in the longest song performed by any animal, a tune that can go on for nearly 24 hours at a time. [46]

Songs of the Humpback Whale had a huge influence, even helping to inspire a worldwide "Save the Whale" movement—a theme that has sci-fi echoes in the 1986 movie *Star Trek IV: The Voyage Home*. The film's highly improbable plot sees an alien space probe coming to Earth in search of humpback whales, replicating their songs in an attempt to communicate with them. But the species has been hunted to extinction by *Star Trek*'s 23rd century, thus providing the perfect excuse for a time travel story.

The same year as *Star Trek IV*, 1986, saw the publication of Arthur C. Clarke's novel *The Songs of Distant Earth*, which also includes a brief mention of whale song—in a supposed quote from a 35th century composer:

> When I wrote *Lamentations for Atlantis*, I had no specific images in mind; I was concerned only with emotional reactions, not explicit scenes; I wanted the music to convey a sense of mystery, of sadness—of overwhelming loss. I was not trying to paint a sound-portrait of ruined cities full of fish... You know, of course, that I based that theme on the songs of the great whales, those mighty minstrels of the sea with whom we made peace too late. [47]

In reality, the world didn't have to wait until the 35th century for human musicians to latch onto whale song, which has become one of the standard clichés of the New Age genre. But long before that genre existed—and before Arthur C. Clarke wrote *The Songs of Distant Earth*—there was a remarkable piece of whale-inspired music in the form of George Crumb's *Vox Balaenae*, which dates from 1971.

Written for a small ensemble of piano, flute and cello, the title is Latin for "voice of the whale"—and at certain points in the work the cello produces a spookily realistic imitation of whale sounds. But the piece goes much further than that, encompassing the whole history of planet Earth—as pianist Andrew Russo explains:

> After coming across recordings of authentic humpback whale calling, Crumb began to devise ways of incorporating these sounds into a large-scale work. From this starting point, he gradually worked out the three-part structure—prologue, theme and variations, and epilogue—that tells the evolutionary tale of time past, time present and time future. [48]

From the point of view of science (or science fiction) geeks, the resulting work is noteworthy for a couple of other reasons too. The musical variations are named after geological eras—such as Archaeozoic, Proterozoic, Palaeozoic, Mesozoic and Cenozoic—while the prologue includes a humorous parody of the opening bars of Richard Strauss's *Also Sprach Zarathustra*—better known as the "theme tune" of *2001: A Space Odyssey*.[1]

References

1. P. Griffiths, *Modern Music and After* (Oxford University Press, Oxford, 1995), p. 37
2. P. Griffiths, *Modern Music and After* (Oxford University Press, Oxford, 1995), p. 89
3. E. Chang, Gruppen. Sounds in Space (2014). http://stockhausenspace.blogspot.com/2014/12/opus-6-gruppen.html
4. P. Griffiths, *Modern Music and After* (Oxford University Press, Oxford, 1995), p. 59
5. P. Griffiths, *Modern Music and After* (Oxford University Press, Oxford, 1995). p. 79
6. P. Griffiths, *Modern Music and After* (Oxford University Press, Oxford, 1995), pp. 62–63
7. J.J. Coupling [John R. Pierce], *Science for Art's Sake* (Astounding Science Fiction, November 1950), pp. 83–92
8. A. Fraknoi, Interdisciplinary Approaches to Astronomy: The Music of the Spheres. Research Gate, August 2016., https://www.researchgate.net/publication/306746125_Interdisciplinary_Approaches_to_Astronomy_The_Music_of_the_Spheres
9. J. Cott, *Stockhausen: Conversations with the Composer* (Pan Books, London, 1974), p. 70

---

[1] Interestingly, in light of Clarke's fictitious *Lamentations for Atlantis*, the Black Box CD of *Vox Balaenae* includes another piece by Crumb entitled "From the Kingdom of Atlantis, circa 10,000 BC" [47].

10. R. Maconie, *Stockhausen on Music* (Marion Boyars, London, 1989), p. 45

11. Wikipedia article on "Symphony No. 1 (Davies)". https://en.wikipedia.org/wiki/Symphony_No._1_(Davies)

12. C. Wilson, *Peter Maxwell Davies: Sinfonietta Accademica* (CD liner notes, Unicorn, 1989)

13. Wikipedia article on "Magic square". https://en.wikipedia.org/wiki/Magic_square#Magic_squares_in_popular_culture

14. P.M. Davies, *The Lighthouse* (CD liner notes, Naxos, 2014)

15. P. Griffiths, *Peter Maxwell Davies* (Robson Books, London, 1982), p. 164

16. G.E. Roberts, Composing with Numbers: Sir Peter Maxwell Davies and Magic Squares. http://mathcs.holycross.edu/~groberts/Courses/Mont2/homepage.html

17. Institute for Research and Coordination in Acoustics/Music. https://www.ircam.fr/lircam/

18. OpenMusic 6.6 User Manual. http://support.ircam.fr/docs/om/om6-manual/co/OM-User-Manual.html

19. C. Deliège, *Conversations with Pierre Boulez* (Eulenberg Books, London, 1976), pp. 60–61

20. C. Wuorinen, *Simple Composition* (Longman, New Your, 1979), p. vii

21. E. Maor, *Music by the Numbers* (Princeton University Press, Princeton, 2019), p. 125

22. B.M. Stableford, *Science Fact and Science Fiction* (Routledge, New York, 2006), p. 315

23. T.M. Tonietti, Albert Einstein and Arnold Schoenberg correspondence (1997). https://doi.org/10.1007/BF02913641

24. E. Maor, *Music by the Numbers* (Princeton, Princeton University Press, 2019), p. 129

25. Arnold Schoenberg, on First Visit, Receives the American Press. Arnold Schoenberg Centre, https://www.schoenberg.at/index.php/en/1933-r-p-arnold-schoenberg-receives-the-american-press

26. Nine of the Best Pieces of Music Inspired by Science. Classical-music.com, July 2016, http://www.classical-music.com/article/nine-best-pieces-music-inspired-science

27. R. Maconie, *Stockhausen on Music* (Marion Boyars, London, 1989), pp. 52–53

28. F. Sherry, *Schoenberg String Quartets Nos 3 and 4* (CD liner notes, Naxos, 2010)

29. J. Cott, *Stockhausen: Conversations with the Composer* (Pan Books, London, 1974), p. 223

30. A. Lewis, *Mantra* (CD liner notes, Naxos, 2010)

31. R. Maconie, *Stockhausen on Music* (Marion Boyars, London, 1989), p. 37

32. J. Cott, *Stockhausen: Conversations with the Composer* (Pan Books, London, 1974), pp. 64–65

33. P. Driver, *The Sunday Times Modern Classics* (CD liner notes, EMI, 1994)

34. P. Griffiths, *Modern Music and After* (Oxford University Press, Oxford, 1995), p. 91

35. Wikipedia article on "Formalized music". https://en.wikipedia.org/wiki/Formalized_Music
36. J.R. Pierce, *The Science of Musical Sound* (Scientific American Books, New York, 1983), p. 220
37. G. Ligeti, *Works for Piano* (CD liner notes, Sony, 1996)
38. P. Glass, *Kepler* (DVD, Orange Mountain Music, 2001)
39. J. Rodgers, W. Ruff, Kepler's Harmony of the World: A Realization for the Ear. Am. Sci. May–June 1979, https://www.jstor.org/stable/27849220
40. F. Terenzi, *Music From the Galaxies.* http://www.fiorella.com/projects.html#galaxies
41. F. Terenzi, *Acoustic Astronomy: The Sounds of the Universe.* Florida International University. http://faculty.fiu.edu/~fterenzi/research/
42. "Fiorella Terenzi", Florida International University. https://case.fiu.edu/about/directory/profiles/terenzi-fiorella.html
43. A.H. Parker, *Supernova Sonata.* http://www.astro.uvic.ca/~alexhp/new/supernova_sonata.html
44. J. McEnery, *The Sound of a Fermi Gamma-Ray Burst.* https://blogs.nasa.gov/GLAST/2012/06/21/post_1340301006610/
45. C.A. Droppelmann, R.E. Mennickent, *Creating Music Based on Quantitative Data from Variable Stars.* https://arxiv.org/abs/1811.02930
46. D. Rothenberg, *Nature's Greatest Hit: The Old and New Songs of the Humpback Whale.* Wire, September 2014., https://www.thewire.co.uk/in-writing/essays/nature_s-greatest-hit_the-old-and-new-songs-of-the-humpback-whale
47. A.C. Clarke, *The Songs of Distant Earth* (Harper-Collins, London, 1994), p. 214
48. A. Russo, *Voice of the Whale* (CD liner notes, Black Box Music, 2002)

Science Fiction and Music Culture

This chapter broaches a huge subject, and can only touch on a few of the ways in which music culture was influenced by science fiction in the last decades of the 20th century. We take a brief look at some of the rock musicians who took inspiration from SF—and, maybe less well known, a couple of operas too. Then there's the fascinating, if improbable-sounding, subject of those musicians, from Stockhausen to Sun Ra, who claimed to have an extraterrestrial connection (or, in the case of Rosemary Brown, a ghostly one). Finally we look at a handful of SF authors, such as Michael Moorcock and Somtow Sucharitkul, who have been equally productive in the music field.

SF-Inspired Music

To many people the phrase "sci-fi music" means songs with science-fictional titles and/or lyrics. This book, on the other hand, is more about science-fictional *sounds* in music. As we saw in the first chapter, such sounds often featured in movies of the 1950s and 60s, but with their origins in the classical avant-garde they're hardly the catchy kind of sounds that are going to make it into a pop song—even one that's explicitly inspired by science fiction.

It also has to be admitted that few sci-fi song lyrics—at least in the 20th century, which is what this book is about—are on a par with SF greats like Robert A. Heinlein, Arthur C. Clarke or Isaac Asimov. That's no surprise, really, because just as a song's tune has to appeal to an enormously wide audience, so do the lyrics. The need to be relatable and relevant to the listener tends to militate against genuine SF themes.

Take David Bowie's "Life on Mars" (1971), for example, which is arguably one of the greatest pop songs of all time. It's certainly one of the most musically sophisticated, with a recent BBC documentary describing it as "a revolutionary song, in which classical music collides with pop brilliance" [1]. But the plain fact is that it's not about life on Mars. It's about life on Earth—and pretty mundane life at that.

In a similar way, the song "Starship Trooper" (1971) by Yes has very little in common with Heinlein's award-winning book *Starship Troopers* (1959). Pink Floyd's "Childhood's End" (1972) shares its title with Arthur C. Clarke's 1953 novel, and almost nothing else. The album *I Robot* (1977) by the Alan Parsons Project has only the most tenuous connection with Asimov's (differently punctuated) collection *I, Robot* (1950). In all these cases, the SF works may have provided the musicians with their initial inspiration, but after that they went off in a different direction of their own.

What's more significant than the occasional borrowing of a title is the way sci-fi tropes were absorbed into the stage acts and personas of musicians like David Bowie and Pink Floyd in the 1960s and 70s. That's a subject we'll come back to later in this chapter, along with such phenomena as the "space jazz" of Sun Ra and the "space rock" of Hawkwind.

Staying on the subject of SF-inspired lyrics for the moment, one song that is completely on-topic for this book—dealing as it does with the three-way interface between science, sci-fi and music—is Kate Bush's "Experiment IV" from 1986.[1] Like many of her songs—but unlike the great majority of hit singles—it presents a first-person narrative from the point of view of a fictional character who clearly isn't the singer/songwriter herself. In the case of "Experiment IV", the story deals with an attempt to build an acoustic weapon. "We were working secretly for the military," it begins, before going on: "It was music we were making here until they told us all they wanted was a sound that could kill" [2].

Writing in the *Guardian*, Alexis Petridis suggested the song might be "a kind of allegory for Bush's perfectionist approach to recording and belief in music's power" [3]. Whatever the case, it's a dramatic idea, reminiscent of the use of Wagner's "Ride of the Valkyries" as a psychological weapon in the film *Apocalypse Now* (1979). In fact there really are such things as sonic weapons—but only for non-lethal purposes like crowd control, not for killing people as in Bush's dystopian vision.

---

[1] I'm grateful to David Hambling for drawing this to my attention.

If things had turned out differently, the 1970s might have seen a crossover between sci-fi and rock by two of the biggest names in each field. To quote the Wikipedia entry on Isaac Asimov, "In December 1974, former Beatle Paul McCartney approached Asimov and asked him if he could write the screenplay for a science-fiction movie musical" [4]. Asimov described the situation in his autobiography *In Joy Still Felt*:

> He had the basic idea for the fantasy, which involved two sets of musical groups—a real one, and a group of extraterrestrial impostors. The real one would be in pursuit of the impostors and would eventually defeat them, despite the fact that the latter had supernormal powers. [5]

Unfortunately, although Asimov was forthcoming with a plot synopsis, it wasn't the sort of thing McCartney was looking for, so the project never went ahead.

A few serious attempts have been made to produce long-form musical adaptations of pre-existing SF stories. One of the earliest and most successful was Rick Wakeman's album *Journey to the Centre of the Earth* (1974), based on Jules Verne's novel of the same name and incorporating spoken narrative from the original text. A crossover rock/classical work, it features the London Symphony Orchestra and the English Chamber Choir in addition to rock musicians—with Wakeman himself playing mellotron, Moog synthesizer and several other keyboard instruments.

A few years later, in a similar vein, came Jeff Wayne's musical version of *The War of the Worlds* (1978), based on the novel by H. G. Wells. As with Verne's novel, this dates from the 19th century, but it seems odd that a similar treatment wasn't accorded to more recent SF classics. Perhaps the closest was Mike Oldfield's album *The Songs of Distant Earth* from 1994, based on the novel by Arthur C. Clarke published just eight years earlier. Oldfield's album doesn't include any narrative, but the section titles do bear an approximate correspondence to the novel—and the result was endorsed by Clarke himself:

> Since the finale of the novel is a musical concert, I was delighted when Mike Oldfield told me that he wished to compose a suite inspired by it… and now, having played the CD-ROM of *The Songs of Distant Earth*, I feel he has lived up to my expectations. [6]

In the classical world, SF-inspired music is even rarer than in popular music. Looking at the standard opera repertoire—i.e. operas that are constantly being restaged in different productions—there's only one that qualifies as serious science fiction. This is *The Makropulos Case*, written in 1926 by the Czech composer Leoš Janáček. It's based on a stage play written a few years earlier by another Czech, Karel Čapek—the man who gave the world the word "robot" in one of his other plays, *R.U.R.*

The central character of *The Makropulos Case* is a young woman named Emilia Marty, who takes an interest in a legal case that has been running for a century—and who talks about events of the 1820s as if she had been there. It gradually emerges that she's lived for hundreds of years, using various different names all with the initials EM. This takes us back to what, from the point of view of English-speaking readers, would be called the Elizabethan age. One of the most colourful characters of that time was Queen Elizabeth's astrologer, John Dee—who also had a Czech connection, because another of his patrons was Prague-based Emperor Rudolf II. The latter just happened to be obsessed with alchemy and the search for eternal life.

In the fictional world of *The Makropulos Case*, an alchemist named Hieronymus Makropulos is supposed to have offered Rudolf a life-extending potion, which the Emperor insisted should first be tested on the alchemist's daughter, Elina Makropulos—Emilia's real identity. The potion put her into a coma, which put the Emperor off—but eventually she woke up to find that she had stopped aging.

Čapek's original play was a comedy, but Janáček turned it into a serious work of science fiction—which may be why the opera is now much better known than the play. As Tim Ashley wrote in the *Guardian*:

> One of classical music's great existential statements, *The Makropulos Case* is phantasmagoria as well as psychodrama. Emilia, victim of a scientific experiment that has monstrously prolonged her existence, is forced to confront the fact that life has meaning precisely because it is finite. [7]

Another opera that's well worth mentioning at this point is *Le Grand Macabre* (1977) by György Ligeti—a composer who appeared prominently in the first chapter due to the use of some of his early experimental works in *2001: A Space Odyssey*. By the time of *Le Grand Macabre*, however, Ligeti's musical style had become much more "normal", and the opera is more striking for its outrageous plot than anything else. In the light of Ligeti's rather po-faced comments about Kubrick's *2001* ("a piece of Hollywood shit"), it's surprising to find him writing what in places is pure slapstick comedy.

The result is one of Ligeti's most popular works (Fig. 5.1)—and, according to the DVD booklet, "one of the most successful operas written in the second

Fig. 5.1 Scene from a 2009 production of Ligeti's comically surreal science fiction opera, *Le Grand Macabre* (Wikimedia Commons user Brustige-file, CC-BY-SA-4.0)

half of the 20th century in terms of the number of productions" [8]. The same booklet goes on to summarize the story as follows:

> The basic plot is concerned with the sudden appearance of Nekrotzar, the Grand Macabre of the title, in a fictional Breughelland (based upon the paintings of the Flem Pieter Brueghel the Elder), his announcement of the end of the world, and its consequences for the characters. The main themes of the opera include false prophets, corrupt government, an incompetent and insane intelligence service and a hedonistic final moral. [8]

As with popular music—think of David Bowie's "Life on Mars"—classical music has its share of works that sound like they ought to be sci-fi but aren't. The most notorious example is Gustav Holst's orchestral suite *The Planets* (1918), which many people imagine to be about astronomy and space travel. But in reality it's nothing of the kind, being based instead on the mediaeval mumbo-jumbo found in the horoscope sections of tabloid newspapers.

Another classical work that sounds like it ought to be SF is the opera *Doctor Atomic* (2005) by John Adams. It's actually about the testing of the first atomic bomb in 1945, centring around the controversial physicist Robert Oppenheimer—the "Doctor Atomic" of the title. Nevertheless, this title is a deliberate reference to old-style B-movie sci-fi—a link that's also reflected in the music. As Thomas May writes in the DVD booklet:

Another component of the score's sonic architecture involves the computer-controlled sounds we hear in the opening of the prelude, which again come to the fore in denser layers in the opera's final minutes. This electronic collage of alien, non-human, metallic shards of noise—inspired by the music of radical-modernist composer Edgard Varèse and suggesting, as Adams recalls, "a post-nuclear holocaust landscape"—was, together with science fiction movie music of the 1950s, the first musical impulse Adams had for the opera. [9]

Doctor Atomic isn't the only opera to draw on science fiction tropes. Michael Tippett's *New Year* (1989) opens with the arrival of an alien spaceship, to the accompaniment of typical electronic "sci-fi" sounds. In Tippett's own words:

For the opera *New Year*, certain special effects were needed, some of which might (had I been living in an earlier century) have been created by standard orchestral means: flight music for a spaceship. I felt, however, that electro-acoustic realization would enhance the distinction between the two worlds of the opera, "Somewhere Today" and "Nowhere Tomorrow". [10]

The spaceship is really only incidental to the opera, which is more in the genre of a futuristic dystopia. The music incorporates an interesting mix of classical and modern sounds, as media theorist Andrew Burn explains:

The drama takes place on New Year's Eve in Terror Town, an imaginary city "Somewhere Today", where with the help of space voyagers from "Nowhere Tomorrow" the principal characters face up to life in a violent, blighted society… Emphasizing saxophones, electric guitars and a large battery of percussion, the scoring often has a bright, raw sound. A pre-recorded tape evokes the spacecraft, while ethereal taped voices are heard in "Donny's Dream". [11]

Extraterrestrial visitors also make an appearance in another classical work, *Sirius* (1977), by a composer who's cropped up several times in this book already—Karlheinz Stockhausen. In the words of musicologist Paul Griffiths, *Sirius* is "a 90-minute enactment in which four virtuosos—a soprano, a bass, a trumpeter and a bass clarinettist—arrive from outer space to instruct the inhabitants of Earth, against a continuous electronic soundtrack" [12].

The title "Sirius" was far from being a random choice. The fact is, according to Stockhausen, he might actually have originated there. In his words:

I have dreamt several times that I came from Sirius and that I was trained there as a musician; it was almost like an obsession during three, four years, and I began to collect information and compose electronic music which was called *Sirius*. [13]

This brings us on to a whole new subject—and one of the strangest aspects of music culture in the latter half of the 20th century: that of musicians who claimed (and/or appeared) to come from outer space.

Extraterrestrial Musicians

With regard to Stockhausen's alien credentials, here is what Tom Service had to say on the subject in a *Guardian* article in October 2005:

Composer Karlheinz Stockhausen has turned himself into a musical myth. This is the man who has influenced everyone from Brian Eno to Björk, and who appeared on the cover of the Beatles' *Sergeant Pepper* album, sandwiched between Carl Jung and Mae West… As if that wasn't enough, this musical pioneer has claimed that he comes not from Burg Mödrath, near Cologne (listed as his birthplace on his biography), but rather from a planet orbiting the star Sirius, and that he was put on Earth to give voice to a cosmic music that will change the world. He is, to put it mildly, a one-off. [14]

To be fair to Stockhausen, he probably didn't really believe he came from Sirius, but was just struggling to express the essential otherworldliness of his music. That becomes clearer in another quote from him in the same *Guardian* article: "Whenever my music or a moment in my music transports a listener into the beyond—transcending time and space—they experience cosmic dimensions" [14].

Musical inspiration often appears in a composer's head without them knowing where it comes from, and the way they react to it will depend on their personality. A religious person might attribute it to a deity, while someone like Stockhausen might feel it came from outer space. Even so, he recognized the subjective nature of this, saying on another occasion that "a memory stored in the psyche is sometimes taken for a cosmic influence" [15].

Another way that intuition can be perceived—which is closer to supernatural fiction than sci-fi—is to attribute it to the spirits of dead people. That's something Stockhausen also talked about:

I've once in a while had the physical experience in which a dead composer was standing behind me while I was working. This has happened to me with Schoenberg, Webern and also Bach and Beethoven, and Mozart. I can't explain this, and people will think I'm insane, but when I've been very tired or when I was struck with a problem, then somehow I received help. [16]

While Stockhausen may have received occasional prompts from deceased composers, another person—the medium Rosemary Brown (1916–2001)—made a whole career out of liaising with them (Fig. 5.2).

Here is Alan Murdie writing on the subject in *Fortean Times* in 2018:

From 1964 Rosemary Brown claimed to be channelling original musical compositions from the post-mortem personalities of Liszt, Beethoven, Bach, Brahms, Schumann, Schubert to Debussy, Rachmaninov, Stravinsky, Gershwin, and (after 1980) John Lennon… The music she created seemed well beyond her

Fig. 5.2 The medium Rosemary Brown, who claimed to transcribe posthumous music by Beethoven, Liszt, Rachmaninov and many other composers (Louis-Maxime-Dubois, CC-BY-SA-4.0)

limited musical knowledge and talents and exactly how it was produced has never been explained. [17]

Her first encounter was with Franz Liszt, probably the greatest composer of piano music of the 19th century. She became aware of his presence while she was improvising at the piano, and felt him guiding her hands over the keyboard. He even spoke to her (having posthumously learned idiomatic English), and told her the name of each composition as she transcribed it onto paper. To quote from the Murdie article again:

There were sceptics aplenty, and newspaper music critics wagged their heads. For example, Dennis Matthews in *The Listener* described her as "delightfully frank and humble", but raised a number of technical reasons, such as mistakes in the niceties of her musical grammar, suggesting that the music came from her own subconscious self. The general view was that the music was notably inferior to the best produced by the various composers when they were on Earth. [17]

Nevertheless she had her admirers. Regarding one of the "Liszt" compositions, the famous Liszt expert Humphrey Searle wrote that "it could well be something that he would have written had he lived another two years"—while the composer Richard Rodney Bennet said "you couldn't fake this music" [17].

In due course Rosemary Brown became something of a star in her own right, as Stuart Jeffries explained in the *Guardian* in December 2019:

She appeared on Oscar Peterson's TV programme and the Johnny Carson Show. Leonard Bernstein dined her at the Savoy and then played some of Brown's transcriptions, being especially thrilled by her Rachmaninov. Colin Davis asked her to inquire of Berlioz about the tempi in *Les Troyens*. [18]

This was in the 1970s, which—shifting the focus from classical music to rock—also happened to be a peak period for extraterrestrial musicians. Take Brian Eno, for example, who featured in the chapter on "The Electronic Revolution". As Jason Heller writes in his book *Strange Stars*, "early in his career, Eno asserted that he'd travelled to Earth from his true home, a planet called Xenon" [19].

Also worth mentioning is the Canadian band Klaatu, formed in 1973 and briefly rumoured to be the Beatles under a new name. Of course they weren't—but was their most famous song, "Calling Occupants of Interplanetary Craft" (better known in a cover version by the Carpenters), a clue to their true identity? As Andy Roberts wrote in *Fortean Times* in 1996:

Fig. 5.3 Sun Ra, allegedly from Saturn, performing on stage in 1992 (Pandelis Karayorgis, CC-BY-SA-2.5)

If Klaatu weren't the Beatles who were they? Well, numerologist John Squires studied the hieroglyphs on their *Hope* album and determined that the band were in fact extraterrestrials. [20]

The most famous "extraterrestrial" musician of all, however, was the prolific jazz composer and keyboardist Sun Ra (see Fig. 5.3).

The strange tale of Sun Ra's origin was recounted by Ian Simmons in a 2009 issue of *Fortean Times*:

Born 22 May 1914 in Birmingham, Alabama, he abandoned his birth name of Herman Blount and took on the name and persona of Sun Ra—Ra being the ancient Egyptian god of the Sun. He claimed that he was of the "angel race" and came not from Earth, but Saturn. This revelation of his true identity seems to have come to him following a visionary experience he claimed to have had while still at college in 1936 or 1937. While involved in deep religious contemplation, Ra said that a bright light had appeared around him, then: "My whole body changed into something else… I landed on a planet that I identified as Saturn; they teleported me down and I was down on stage with them. They wanted to talk with me. The had a little antenna over each ear. A little antenna over each eye. They talked to me. They told me to stop attending college… I would speak through music, and the world would listen. That's what they told me." [13]

As far-fetched as the story is, Ra's popularity and larger-than-life personality were such that many people were happy to go along with it—as the following anecdote from the Simmons piece shows:

> Late in life, when admitted to hospital he insisted on having Saturn entered as his place of birth on the admission form. Nurses summoned a psychiatrist who responded: "This is Sun Ra—of course he's from Saturn!" [13]

Sun Ra's music was as out-of-this-world as his persona. The peak of his creativity, in the late 1950s and 1960s, coincided with new developments in jazz that were, in their own way, just as experimental as the work of Stockhausen and Boulez on the classical side. Sun Ra was at the cutting edge of this avant-garde jazz sound, which came across particularly strongly in his live stage shows. These involved a varying group of supporting musicians he referred to as the "Arkestra", as Simmons explains:

> It was a vast and sprawling ensemble through which huge numbers of musicians passed, dressed in flamboyant costumes and headdresses in which ancient Egypt and science fiction collided. And if the Arkestra didn't look like your average jazz combo, then the music they produced was equally unique and constantly surprising—a mixture of lengthy, avant-garde jazz work-outs… and, in later years, Ra's own Moog synthesizer solos and chants about space travel. [13]

There had never been anyone in music quite like Sun Ra—who, as philosophy professor Jerome Langguth points out, managed to be simultaneously cosmic and comic:

> Sun Ra's art is the art of the sublime in its persistent interest in the cosmic, the unrepresentable and the impossible. At the same time Ra's art and music, as well as his persona, are resolutely comic. Ra routinely combined the experimental with the accessible, the humorous, the traditional and even the deliberately cheesy. [21]

As befits a native of Saturn, Ra's whole output exudes a distinctly spacey, futuristic feel—which comes across in the titles he chose for his works as much as anything else. To quote Langguth again:

In addition to the costumes and space chants, Sun Ra's recorded works typically bore titles that reflected Ra's debt to science fiction. To name just a handful: *We Travel The Space Ways* (1956), *Sun Ra Visits Planet Earth* (1956), *Cosmic Tones for Mental Therapy* (1962), *The Heliocentric Worlds of Sun Ra* (1965), *Outer Spaceways Incorporated* (1966) and *Monorails and Satellites* (1966).

He then adds that:

In 1972, Ra also collaborated with director John Coney on *Space Is the Place*, a film starring Sun Ra and the Arkestra as themselves, which owed much visually, sonically and thematically to the B science fiction films of the 1950s. [21]

The movie has a complicated plot, to say least—but here is Langguth's brief summary of it:

Sun Ra, who is travelling in a spacecraft powered by the sound of the Intergalactic Solar Myth Arkestra in full force, locates a planet of great beauty and mystery that he deems an appropriate future home for oppressed African Americans. Ra will bring them there, he intones, through "isotope teleportation, transmolecularization, or, better still, transport the whole planet here through music." Ra and the Arkestra then return to Earth, apparently in order to save African Americans from the debilitating realities of racist and materialist American culture.

Once on Earth, Ra fights a kind of metaphysical pimp known as the Overseer, while at the same time he is interfered with by the FBI, and possibly NASA as well… In the Earth scenes, Ra sets up an operation called the "Outer Spaceways Incorporated Employment Agency" designed to help people to realize their "alter-destiny"… and successfully rescues at least a handful of people before returning to space. As the Arkestra's craft disembarks, the Earth is blown apart as the band perform "Space is the Place," one of Ra's most memorable space chants. [21]

Unlike Sun Ra, David Bowie never said he came from outer space, leaving that to his outrageous on-stage alter-ego Ziggy Stardust instead. Even so, there were times when people could be forgiven for thinking Bowie was an alien. As Dean Ballinger wrote in *Fortean Times* shortly after the singer's death in 2016:

In the 1970s Bowie quickly became established as an icon of alienness thanks to his unusual appearance, space-themed tunes and otherworldly personae, notably Ziggy Stardust. It was an identification cemented by his starring role in Nicolas Roeg's arty 1976 sci-fi flick *The Man Who Fell To Earth*. [22]

In fact Bowie established his SF credentials even earlier, with the release of the hit single "Space Oddity" in July 1969—the same month as the Apollo 11 lunar landing. It was the song that, according to Jason Heller, "launched sci-fi music in earnest":

The song didn't just contain sci-fi lyrics. Sonically, it was a reflection of sci-fi, full of futuristic tones and the innovative manipulation of studio gadgetry. [19]

Interestingly, despite the song's proximity to the real-world Moon landing, its primary inspiration came from science fiction—as Jack Needham explains on the BBC website:

The song was actually informed by Bowie's fascination with space as a whole, especially influenced by Stanley Kubrick's 1968 film *2001: A Space Odyssey*. "In England, it was always presumed that it was written about the space landing, because it kind of came to prominence around the same time. But it actually wasn't," he later said. "It was written because of going to see the film *2001*, which I found amazing… It was really a revelation to me." [23]

To state the obvious, the 1960s was an extraordinary decade. It saw both the first humans in space and the first to walk on the Moon. It saw a work of deeply philosophical science fiction, in the form of Kubrick's *2001*, reach a mass audience around the world. With artists like Frank Zappa, the Beatles and Bowie himself, it saw popular music developing a range and depth previously only found in classical music.

At the same time, the sixties was a crazy, mixed-up decade. Interests that had previously been confined to the esoteric fringe—such as Zen Buddhism, the paranormal and ufology—entered the mainstream. In some cases, at least, a certain chemical had a lot to answer for: the psychedelic drug lysergic acid diethylamide, better known as LSD or simply "acid".

The link between LSD and UFOs was particularly inseparable, as cultural historian S. D. Tucker points out:

When LSD first hit the streets of swinging London in 1965, one especially potent batch was known as 'flying saucers', while one of the hippest clubs of the day was called UFO. [24]

There's another three-way intersection of music, science and sci-fi here, because John Hopkins, one of the founders of the UFO club on London's

Tottenham Court Road, was "a nuclear physicist turned photojournalist", according to media historian Gary Lachman [25].

One of the UFO club's resident bands, Pink Floyd, went on to became as big a name as David Bowie—and just as instrumental in bringing sci-fi imagery into the world of rock music. In Pink Floyd's case, there was a distinctly psychedelic dimension too, as social historians David Clarke and Andy Roberts explain:

> Pink Floyd's first album *The Piper At The Gates Of Dawn* included the atmospheric paean to deep space, "Astronomy Domine", possibly the first song to use outer space as a metaphor for inner space. By their second album Pink Floyd had further absorbed saucer culture, entitling it *A Saucerful of Secrets*, mixing ideas of flying saucers, the secrets found inside the mind, with perhaps a nod toward a batch of potent LSD called "flying saucers". The sleeve artwork left fans in no doubt that space, inner or outer, was the place; swirling universes and spinning discs mixed with signs of the zodiac, and the keynote song, "Set the Controls for the Heart of the Sun", became the backdrop for many psychedelic journeys toward dawn. [26]

The band's unusual name is normally interpreted as a homage to two blues musicians, Pink Anderson and Floyd Council—but there's another, more out-of-this-world possibility. In the words of Gary Lachman:

> After group founder Syd Barrett spent a year living in a flat in which virtually everything was spiked with acid, he claimed the name had been transmitted to him by an extraterrestrial. [25]

Like Bowie, Syd Barrett (Fig. 5.4) was a fan of science fiction, and both the lyrics and the sound of his songs were thoroughly infused with the genre. To quote Jason Heller:

> Barrett had absorbed cosmic reverberations and modulated echoes from space-themed music as a sci-fi-loving youth in the late '50s and early '60s; he also came of age gripping copies of the sci-fi comic book *Dan Dare*, as well as novels like *Childhood's End* by Arthur C. Clarke. [19]

Sadly, Barrett's tenure with Pink Floyd was short-lived. His over-indulgence in hallucinogenic drugs had a damaging effect on his mental health, and he was dropped from the band during the recording of *A Saucerful of Secrets* in 1968.

Fig. 5.4 Syd Barrett, the founder member of Pink Floyd responsible for songs like "Astronomy Domine" and "Interstellar Overdrive", who once claimed the group's name was given to him by an alien (Wikimedia Commons user Bojars, CC-BY-SA-3.0)

Following Barrett's departure, the explicitly "spacey" content of Pink Floyd's music dropped almost to zero, and the band's later albums like *Dark Side of the Moon* (1973), *Animals* (1977) and *The Wall* (1979) are predominantly socio-political in their concerns. It's a measure of Barrett's influence, however, that even to this day Pink Floyd's music is often pigeonholed as "space rock".

Other bands, however, continued to embrace the genre wholeheartedly—none more so than Hawkwind. Here's what Ian Simmons has to say about the 1973 album *Space Ritual*, based on material from one of Hawkwind's spectacular live shows:

> According to band member Nik Turner, the album concerns seven cosmonauts who are travelling through space in a state of suspended animation and is essentially an audiovisual portrayal of their fantasies and dreams on the voyage. [13]

It's a fascinating premise, which could have been turned into a serious science fiction story. Instead, Turner's narrative is a humorous send-up of the genre's glibbest and most far-fetched conventions. The tone can be gathered from the following excerpt from his liner notes:

> Coordinates centring at red delta two plus seven. The witches hurriedly led off the sorceress. With the ease of a band that had made many such landings, the men went efficiently to their tasks. All was quiet in the chamber except for the monotonous printout of the robo-screen. "Astrological conditions favourable. Terran data system functioning."
>
> The planet jerking and jumping in the visor globe began to show brown and green with an occasional flash of blue. Nik adjusted his silver spurs and loosened the thongs of his sword… "Ley lines observed and show information accurate. Twelve minutes to landing." [27]

That's about as far as it's possible to get from the serious literary SF that was being produced in the 1970s. Yet *Space Ritual* has a connection with the latter too, through the involvement of SF author Michael Moorcock. He wrote the lyrics to one of the album's songs, "Sonic Attack"—which coincidentally touches on a similar theme to Kate Bush's "Experiment IV" mentioned earlier—while another track, "Black Corridor", includes verbatim quotes from Moorcock's 1969 novel of that title.

Although Moorcock's contributions to *Space Ritual* were literary rather than musical, he did go on to develop a secondary career as a musician himself—and that brings us on to the subject of the next section.

Musician-Authors

Staying with Michael Moorcock for the moment, he's undoubtedly known best as a prolific author of fantasy and science fiction, with over a hundred novels to his credit. During the 1960s he was also a highly influential editor, taking the helm of Britain's premier SF magazine, *New Worlds*, when he was just 24 (see Fig. 5.5). Under Moorcock's leadership the magazine acquired a prestigious Arts Council grant in 1967—and then in the following year gained the even greater distinction of being, as author Brian Aldiss later recalled, "referred to as degenerate in the House of Commons" [28].

By the 1970s, Moorcock's writing career was supplemented by a sideline in music—and the two were so closely intertwined they can't always be separated. Alongside his contributions to Hawkwind, he also formed a band of his own, called "The Deep Fix"—taken from the title of a novella he'd written back in 1966.

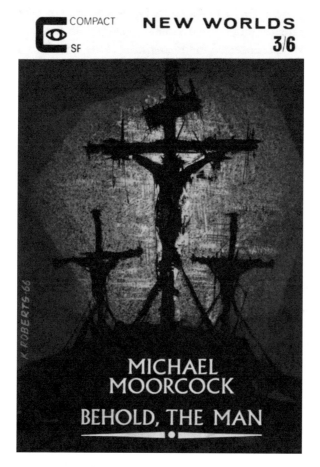

Fig. 5.5 The September 1966 issue of *New Worlds* magazine, edited by Michael Moorcock and featuring his novella "Behold, the Man" (Public domain image)

The story itself isn't about music but hallucinogenics—a subject which, as we saw a few pages ago, was closely related to the music scene in the 1960s. Specifically, "The Deep Fix" deals with therapeutic devices that get out of control: "by different methods—light, sound-waves, simulated brain waves and so on—the machines created the symptoms of dozens of basic abnormalities and thousands of permutations" [29].

The Deep Fix—the band, not the novella—later appeared in another of Moorcock's stories, "A Dead Singer". This dates from 1974, by which time Moorcock's band of that name really did exist—so the intertwining of fiction and reality is taken a step further. A sad fact that's common to both the real world and the fictional one is that rock legend Jimi Hendrix died at the tragically early age of 27 in 1970. In "A Dead Singer", however, the protagonist—a permanently stoned ex-roadie—imagines he has (or just possibly really does have) a resurrected Hendrix riding along with him in his van.

The story name-drops a few other real-world acts, including David Bowie and Hawkwind—and, naturally enough, the Deep Fix. At one point the roadie tells a hitchhiker they were one of the bands he used to drive round—to which the passenger replies "that's a really good band, very heavy" [30].

As an aside, there's a tenuous sci-fi link in the title of one of Hendrix's most famous songs, "Purple Haze". In Jason Heller's *Strange Stars*, after mentioning that Hendrix was a sci-fi fan in his youth, he goes on:

> Hendrix also devoured a 1957 novel by Philip José Farmer titled *Night of Light*. In it, a planet orbiting a distant sun is inundated in a mysterious radiation that causes reality to distort. At one point in the story, the sunspots visible from an alien planet are described as having a "purplish haze". Backstage before a concert in 1966, Hendrix began filling pages with a story that had appeared in his head. He later whittled down that tangle of poetry and used it as the core of a new song. His original elements of sci-fi—including, among other things, "the history of the wars on Neptune"—were left on the cutting-room floor. But the song's title retained that vivid, two-word phrase that had jumped out at Hendrix while reading Farmer's *Night of Light*. Hendrix called the song "Purple Haze." [19]

One of the regular contributors to *New* Worlds under Moorcock's editorship was the author Langdon Jones—and he too had a double life as a musician as well as a writer. He's credited, in fact, as the pianist on one the Deep Fix's more recent albums, *The Entropy Tango & Gloriana Demo Sessions* [31].

For the most part, however, Jones's musical interests lie more on the classical side, both as a listener and as a composer. Writing on his website circa 2006, he said:

> My principal interest these days apart from the internet and local politics is classical music. I have had a number of works performed locally, including an hour-long setting of Mervyn Peake's poem, *The Rhyme of the Flying Bomb*… I have more or less stopped performing and composing, but still enjoy listening and creating MIDI transcriptions. Favourite composers include Schoenberg, Nielsen, Messiaen, Brian, Alkan and Honegger. [32]

Of those names, Schoenberg has already featured multiple times in this book. The French composer Olivier Messiaen (1908–92) was also mentioned briefly in the first chapter, in the context of his use of an electronic instrument, the Ondes Martenot, in his *Turangalîla-Symphonie* of 1949. That's quite a popular work, but Messiaen's music subsequently became "far less accessible", as the liner notes of one CD say of his 1960 composition *Chronochromie*:

> The title *Chronochromie* is taken from two Greek words, to mean "the colour of time", and the composer insisted the orchestral colours are there to bring out the rhythmic and temporal structures. [33]

Messiaen's name is mentioned several times in Jones's short story collection *The Eye of the Lens* (1980). One of the stories was originally constructed around another of Messiaen's work, *Trois Petites Liturgies*—but the composer "refused permission to use his text on various, quite reasonable, grounds" [34]. Another story in the collection, "The Time Machine", makes specific reference to *Chronochromie*:

> The time machine utilises certain objects for its various operations: a skull cap with electrodes… the miniature score of Messiaen's *Chronochromie*, a magnetic tape 2,400 feet long containing nothing but the voice of a man repeating the word "time", a reproduction of Dali's landscape *Persistence of Memory*, a bracket clock. [35]

The most significant use of music in Jones's fiction, however, is his short story "The Music Makers", published in *New Worlds* in November 1965. Set in the future, its protagonist is a violinist on Mars who, at the story's climax, hears—or imagines he hears—the music of the long-dead indigenous Martians.

Earlier in the story, he plays a real-world piece—the Violin Concerto by Schoenberg's pupil Alban Berg—and then discusses the philosophy of music with the orchestra's conductor. The latter argues for the "relativity" of music (an issue we'll come back to in the final chapter of this book)—the view that human music is for human ears only, and "will only create its emotional effect under the right conditions".

The violinist takes the opposite view, arguing that it is only the response to music that is human-centric, not the music itself:

> Great music is something absolute, to which we respond with human emotions. What Berg created was something like these dunes, permanent and absolute, and what feelings Berg creates in us are in essence produced no differently than those caused by the dunes. [36]

Langdon Jones is just one of several SF authors with a classical music connection. Another is Lloyd Biggle (1923–2002), described in the *Encyclopedia of Science Fiction* as follows:

> US author and musicologist, with a PhD in musicology from the University of Michigan on the music of the medieval Flemish composer Antoine Brumel. His interest in music and the other arts, perhaps watered down more than necessary in an effort to make such concerns palatable to his readers, can be detected throughout his science fiction. [37]

Some of Biggle's science-fictional speculations on the future of music will be described in the next chapter. Despite his musical education, however, he was—like Michael Moorcock and Langdon Jones—essentially a professional author who had a side interest in music. If we're looking for someone who's literary and musical careers are more closely balanced, we have to turn to Somtow Sucharitkul (Fig. 5.6)—who has been extraordinarily prolific in both fields.

Fig. 5.6 Somtow Sucharitkul, composer, conductor and science fiction author (Public domain image)

Born in Thailand in 1952, Somtow's family moved to England when he was six months old, so his cultural background and education is primarily British. It wasn't just any British education, either. He attended the most prestigious school of all, Eton College—the same as Prince William, Boris Johnson and many other members of the upper echelon of British society.

Later he moved to the United States, where much of his science fiction was published, and then to his native Thailand, where he became director of the Bangkok Opera company. As well as conducting traditional western operas—including a production of Wagner's *Ring* cycle which had "the official blessing of the composer's great-grandson Wolfgang", according to the *New York Times* [38]—he also wrote the words and music for entirely new operas of his own. Notable among these is *Ayodhya* (2006), which attempts to do for the Indian epic *Ramayana* what Wagner did for Germanic mythology in the *Ring*.

In parallel with his musical career, Sucharitkul has written dozens of novels in the SF, fantasy and horror genres—most of them under the byline "S. P. Somtow". In fact he seems to have something of a dual personality, with "Somtow" being the writer and "Sucharitkul" the musician. A CD of his ballet *Kaki* (1997), for example, credits S. P. Somtow for the scenario and Somtow Sucharitkul for the music [39].

Kaki is based on a humorous legend from Buddhist literature about a queen who is so alluring that everyone falls in love with her. From a musical point of view it's a mish-mash of styles, combining the classical ballet sounds of, say, Tchaikovsky's *Swan Lake* with Broadway musicals and—in some of the melodies—traditional Asian music. The scoring is equally eclectic, for full symphony orchestra supplemented by Indian, Japanese and Indonesian instruments. One section even includes hybrid east/west dances, such as the "Cholonaise" (a Chinese polonaise) and "Japañera" (Japanese habañera). As Sucharitkul says:

> I decided to return to the roots of Russian ballet, and write the kind of score that the great practitioners of Russian ballet might have written if they had been subjected to the seductive influence of Asian musical traditions. [39]

An earlier composition by Sucharitkul, *Starscapes* from 1979, has an interesting sci-fi connection—and not just the obvious one of the title. Playing in the violin section of the orchestra at its first performance in Florida was another prolific SF author, Barry Malzberg [40]. He's perhaps best known for his award-winning novel *Beyond Apollo* (1972), but one of his more obscure novels—*Chorale*, dating from 1976—is worth looking at here because it has a musical theme.

It's often said there's no such thing as an original idea in science fiction, but there's nothing quite like the premise of *Chorale*. Countless novels deal with time travel into the past—either to change it or to prevent others from changing it—but the bizarre twist here is that it has to be constantly recreated to prevent it from disappearing. Specifically, actors from the 23rd century "Department of Reconstruction" have to go back and relive key moments in the lives of famous people, as they're described in history books, because otherwise the world would cease to exist.

It's a nutty sounding idea, and it even sounds nutty to the characters in the novel. It's supposedly based on the controversial work of a maverick scientist named Kemper, who was either a genius or a lunatic:

> Kemper had postulated that the past was in flux, was in fact merely an extension of the present. Furthermore, the past would have to be eternally reconstructed by surrogates from the present who would return in time to live out key aspects of the lives of famed historical personages in order to make sure that the present, seated perilously as it was upon the past, did not dissolve. [41]

It's never made clear to the reader if there's any substance to this, or if the whole thing is just a pseudo-scientific hoax. The truth, of course, is that Malzberg just came up with the idea so he could tell the story he wanted to—about the protagonist, Reuter, re-living key moments in the life of Beethoven.

For the most part *Chorale* is a comedy, but it includes some thought-provoking incidents, such as when Reuter—who's familiar with far more advanced music than Beethoven's—gets frustrated when an orchestra complains about the difficulty of "his" Ninth Symphony:

> "It is going to kill us first," someone said.
> "Do not melodramatize."
> "We do not melodramatize, Maestro. We merely essay to tell the truth."
> "Truth is relative," Reuter said, "and the performance standards here are well within your means. By the late 19th century, let alone the 20th, this work will fall into the range of any modestly competent orchestra and soloists."
> "How do you know?"
> "Because I know," Reuter said furiously, "because you must trust the integrity of my vision… The work will be popular and central to the repertory. It will be much in demand and will please everyone almost all of the time."
> "You are sure of this?" the conductor said. His head was bowed in simple awe. "You sound utterly convincing, but nevertheless –"

"It is the truth," Reuter said. "You cannot imagine, any of you, how low your performance standards are as compared to how they will be in a mere 75 years." [41]

Needless to say, this conversation never took place—not least because, by the time in question, Beethoven was almost totally deaf. That's a point that bothers Reuter in the novel, and is one of several inconsistencies that lead him to suspect he's the victim of a hoax.

It's true, however, that Beethoven was one of the first people to appreciate that musical values are constantly evolving, and that standards would be different in the future. Almost 20 years before the premiere of the Ninth Symphony, the violinist Felix Radicati recounted the following anecdote about Beethoven's "Razumovsky" string quartets, which were notoriously difficult for the players of the time:

Beethoven, as the world says, and as I believe, is mad, for these are not music. He submitted them to me in manuscript… and I said to him that he surely did not consider these works to be music? To which he replied, "Oh, they are not for you, but for a later age." [42]

This brings us neatly on to the subject of the final chapter, which deals with science fictional speculations concerning the future of music.

References

1. Our Classical Century, episode 3: 1953–1971. https://www.bbc.co.uk/programmes/m00041tg
2. K. Bush, Experiment IV (single, EMI, 1986)
3. A. Petridis, Kate Bush – every UK single ranked. Guardian, November 2018. https://www.theguardian.com/music/2018/nov/23/kate-bush-every-uk-single-ranked
4. Wikipedia article on "Isaac Asimov". https://en.wikipedia.org/wiki/Isaac_Asimov
5. I. Asimov, *In Joy Still Felt* (Avon, New York, 1980), p. 693
6. A.C. Clarke, *The Songs of Distant Earth* (CD interior notes, Warner Music, 1994)
7. T. Ashley, The Makropulos Case, Guardian, September 2010. https://www.theguardian.com/music/2010/sep/22/the-makropulos-case-review
8. M. Searby, *Le Grand Macabre* (DVD booklet, Arthaus Musik, 2012)
9. T. May, *Doctor Atomic* (DVD booklet, Opus Arte, 2008)
10. M. Bowen, *Tippett on Music* (Clarendon Press, Oxford, 1995), p. 260

11. A. Burn, *Tippett Symphonies* (CD liner notes, Chandos, 2005)
12. P. Griffiths, *Modern Music and After* (Oxford University Press, Oxford, 1995), p. 223
13. I. Simmons, Mothership connections. *Fortean Times* **244**, 30–35 (2009)
14. Tom Service, Beam me up, Stocky. Guardian, October 2005. https://www.the-guardian.com/music/2005/oct/13/classicalmusicandopera
15. R. Maconie, *Stockhausen on Music* (Marion Boyars, London, 1989), p. 136
16. J. Cott, *Stockhausen: Conversations with the Composer* (Pan Books, London, 1974), p. 195
17. A. Murdie, Unfinished symphonies. *Fortean Times* **363**, 16–18 (2018)
18. S. Jeffries, All hail Rosemary Brown. Guardian, December 2019. https://www.theguardian.com/music/2019/dec/05/rosemary-brown-liszt-beethoven-pianist
19. J. Heller, *Strange Stars* (Melville House, 2018, Kindle edition)
20. A. Roberts, Rocking the alien. *Fortean Times* **88**, 34–38 (1996)
21. J.J. Langguth, *Sounds of the Future* (McFarland, North Carolina, 2010), pp. 148–161
22. D. Ballinger, The mage who sold the world. *Fortean Times* **338**, 28–33 (2016)
23. J. Needham, Intergalactic Soundtrack. BBC website, July 2019. https://www.bbc.co.uk/programmes/articles/4HdFLPrM5Rsw2M6Gdspm26d/intergalactic-soundtrack-11-stellar-songs-inspired-by-space
24. S.D. Tucker, *Space Oddities* (Amberley, Stroud, 2017), p. 162
25. G. Lachman, *Turn Off Your Mind* (Sidgwick & Jackson, London, 2001), p. 346
26. D. Clarke, A. Roberts, *Flying Saucerers* (Loughborough, Alternative Albion, 2007), p. 184
27. Hawkwind, *Space Ritual* (CD liner notes, EMI, 2001)
28. B. Aldiss, *Billion Year Spree* (Weidenfeld & Nicholson, London, 1973), p. 298
29. M. Moorcock, The Deep Fix, in *Spaced Out* (Panther, London, 1977), pp. 13–75
30. M. Moorcock, A Dead Singer, in *The Year's Best SF 9* (Futura, London, 1976), pp. 182–200
31. Discogs entry for *Michael Moorcock & The Deep Fix: The Entropy Tango & Gloriana Demo Sessions* (2008). https://www.discogs.com/Michael-Moorcock-The-Deep-Fix-The-Entropy-Tango-Gloriana-Demo-Sessions/release/4029931
32. L. Jones, Personal Page. https://www.langjones.com/personal.html
33. R. Nichols, O. Messiaen, *Quatuor pour le Fin du Temps* (CD liner notes, Warner, 2015)
34. L. Jones, *The Eye of the Lens* (Savoy Books, London, 1980), p. 8
35. L. Jones, *The Eye of the Lens* (Savoy Books, London, 1980), p. 99
36. L. Jones, The Music Makers, in *New Worlds* no. 156, November 1965, pp. 58–66
37. *Encyclopedia of Science Fiction* entry for "Lloyd Biggle, Jr". http://www.sf-encyclopedia.com/entry/biggle_lloyd_jr
38. R. Turnball, Bangkok Opera presents Wagner's ring with a difference. New York Times, April 2006. https://archive.nytimes.com/www.nytimes.com/iht/2006/04/18/arts/18wagn.html

39. S. Sucharitkul, *Kaki: a Ballet in Two Acts* (CD liner notes, Imaginary Records, 1997)
40. M. Ashley, *The Illustrated Book of Science Fiction Lists* (Virgin, London, 1982), p. 153
41. B. Malzberg, *Chorale* (Gateway, 2011, Kindle edition)
42. A. Thayer, *Life of Beethoven* (The Folio Society, London, 2001), p. 213

Speculations on a Musical Theme

Having focused so far on the portrayal of science fiction in music, this final chapter flips things round to examine the way various musical topics have been handled in SF. First there's the question of how music will evolve in future—will it embrace other senses besides hearing, or maybe even exploit extra-sensory perception? One author who had a lifelong passion for music was Philip K, Dick, and we take a look at his treatment of the subject in novels like *Flow My Tears the Policeman Said* and *VALIS*—the latter, incidentally, being the subject of an excellent opera itself. Finally we ask the all-important question: what would extraterrestrials make of Earth music?

Music of the Future

Music, like all the arts, evolves over time, and the idea that it might sound different in the future has been around for a long time. At the end of the last chapter we saw how Beethoven said he wrote his Razumovsky quartets "for a later age" [1]. In that instance he was referring to performance standards, and those quartets hardly sound futuristic today—although one of them does have a vaguely "spacey" connection in that Beethoven said he conceived its slow movement as he "gazed at the stars contemplating the music of the spheres" [2].

More recently, the title *Music of the Future* was used in the 1950s by Desmond Leslie for a compilation of his *musique concrète* compositions, as mentioned in the chapter on "The Electronic Revolution". Those works were

© The Editor(s) (if applicable) and The Author(s),
under exclusive licence to Springer Nature Switzerland AG 2020
A. May, *The Science of Sci-Fi Music*, Science and Fiction,
https://doi.org/10.1007/978-3-030-47833-9_6

indeed prophetic of the future, in that they anticipated the now widespread use of electronics in music.

Amongst musicians in general, however, anticipating the "future"—by more than a few months, anyway—is a long way down their list of priorities. They're more concerned with riding the crest of current trends, which is how people in any field of the arts get to the top of their field. On the other hand, it's a different matter in science fiction—where speculating about the future is what it's all about.

One author who speculated multiple times on the specific topic of the future of music was Lloyd Biggle. As mentioned in the previous chapter, he studied musicology at university prior to becoming a full-time SF author. It was a background he drew on in several stories, such as "The Tunesmith" from 1957.

This deals with the struggles of a composer named Baque,[1] who aspires to write serious music in a highly commercialized future where the only paying work for composers is the churning out of advertising jingles. The story contains some impressively accurate prophecies about the future of music technology, including synthesizer keyboards—which Biggle calls "multichords"—and the use of machines to harmonize melodies (i.e. to decide which chords ought to accompany a tune). Here is Baque musing to himself:

> Compose something. You're not a hack, like the other tunesmiths. You don't punch your melodies out on a harmonizer's keyboard and let a machine harmonize them for you. You're a musician, not a melody monger. Write some music. Write a sonata for multichord. [3]

In a slightly later Biggle short story, "Spare the Rod" from 1958, a music professor considers another question that was hypothetical then but much more pertinent today—that of automated composition:

> You've heard of the composing machines—the music they wrote was perfectly correct, and horribly dull and naive. Then somebody built one with no system at all. What it produced was absolute chaos, but scattered through that chaos were beautiful tonal effects that the machine happened on by accident. It took a great artist to understand those effects and use them as they should have been. Morglitz used that machine in his last compositions. It inspired some of the greatest things he wrote. [4]

---

[1] This is essentially a pun, since it's how many English-speakers pronounce the name "Bach".

"Morglitz" was a fictional composer of Biggle's invention, but at the time the story was written the real world already had someone who was endeavouring to harness "absolute chaos" in the service of music, in the form of Iannis Xenakis. As mentioned in an earlier chapter ("Scientific Music"), Xenakis used such effects to create what he called "stochastic music".

That very term appears in another short story from the same period, James Blish's "A Work of Art", originally published in the July 1956 issue of *Science Fiction Stories*. It's not clear whether Blish knew of Xenakis's work when he referred to "stochastic music" in the story, but there's another real-world composer he was very definitely aware of. That's Richard Strauss, whose tone poem *Also Sprach Zarathustra*—or rather the opening bars of it—became the "theme music" for *2001: A Space Odyssey*.[2]

Set in 2161, "A Work of Art" sees Strauss's memories and personality implanted in a man of that time, in a way that effectively "resurrects" the composer. He then has to read up about all the latest musical developments—of which sophisticated techniques like stochastic music are very much in the minority:

> By far the largest body of work being produced fell into a category misleadingly called "science-music". The term reflected nothing but the titles of the works, which dealt with space flight, time travel, and other subjects of a romantic or an unlikely nature. There was nothing in the least scientific about the music, which consisted of a melange of clichés, imitations of natural sounds, and stylistic tricks. [5]

That's reminiscent of a point made in the previous chapter, that many supposedly "sci-fi" songs of our own time are science-fictional only in a very superficial way. In Blish's story, Strauss ignores all the fancy new trends, and writes an opera in his own style—that of the early 20th century. The opera, however, isn't the "work of art" of the story's title—it's the re-created Strauss himself, the crowning achievement of "psi-sculptor" Barkun Kris.

At the opera's premiere (see Fig. 1), when the music comes to an end, the audience's rapturous applause is less for Strauss than for Kris—who tells him:

> In a moment, when I speak a certain formulation to you, you will realize that your name is Jerom Bosch, born in our century and with a life in it all your own. The superimposed memories which have made you assume the mask, the persona of a great composer will be gone. I tell you this so that you may understand why these people here share your applause with me. The art of psi-sculpture—

---

[2] Over a decade *after* Blish's story was written, it's worth noting.

The world of 2161 was about to hear a newly-completed opera by
Richard Strauss...

Fig. 1 Illustration from the original magazine appearance of "A Work of Art" by
James Blish: "the world of 2161 was about to hear a newly completed opera by Richard
Strauss" (public domain image)

the creation of artificial personalities for aesthetic enjoyment—may never reach
such a pinnacle again. For you should understand that as Jerom Bosch you had
no talent for music at all; indeed, we searched a long time to find a man who
was utterly unable to carry even the simplest tune. Yet we were able to impose
upon such unpromising material not only the personality, but the genius of a
great composer. [5]

One of the commonest ways that "future music" is portrayed in SF is by
extending it to other senses besides hearing—vision in particular. A good
example with a spacey connection can be found in Robert Silverberg's novel
The World Inside (1971), which includes a detailed description of a 24th century concert by a "cosmos group" of audiovisual instruments. It's full of made-
up technobabble—such as the "vibrastar" played by the band's leader,
17-year-old Dillon Chrimes—but that all adds to the futuristic flavour of
the piece:

He brings his hands up for a virtuoso pounce and slams them down on the projectrons. The old headblaster! Moon and Sun and planets and stars come roaring out of his instrument. The whole glittering universe erupts in the hall… The doppler-inverter noodles in with a theme of its own, catching something of the descending fervour of Dillon's stellar patterns. At once the comet-harp over-lays this with a more sensational series of twanging tones that immediately transmute themselves into looping blares of green light. These are seized by the spectrum-rider, who climbs up on top of them and, grinning broadly, skis off toward the ultraviolet in a shower of hissing crispness… Then the incantator enters, portentous, booming, sending reverberations shivering through the walls, heightening the significance of the tonal and astronomical patterns until the convergences become almost unbearably beautiful. [6]

That sounds pretty cool, but when other senses besides sound and vision are brought in, the result veers towards the comic—as in the following excerpt from Aldous Huxley's satirical novel *Brave New World* (1932):

The scent organ was playing a delightfully refreshing Herbal Capriccio—rip-pling arpeggios of thyme and lavender, of rosemary, basil, myrtle, tarragon; a series of daring modulations through the spice keys into ambergris; and a slow return through sandalwood, camphor, cedar and new-mown hay, with occa-sional subtle touches of discord—a whiff of kidney pudding, the faintest suspi-cion of pig's dung—back to the simple aromatics with which the piece began. The final blast of thyme died away; there was a round of applause; the lights went up. [7]

An olfactory musical analogue is wacky enough, but what about a tactile one? That's exactly what Stanisław Lem's recurring character Ijon Tichy encounters in *The Futurological Congress* (1971)—although Tichy is subjected to so many hallucinations in the course of the novel it might not actually be a real experience:

Symington led me to his study. And again something stupid happened. I turned on this desk unit, taking it for a radio. A swarm of glittering fleas came bursting out, covered me from head to foot, tickling everywhere, all over—until, scream-ing and waving my arms, I ran out into the hallway. It was an ordinary feely; by accident I had switched it on in the middle of Kitschekov's *Pruriginous Scherzo*. I really don't understand this new, tactual art form. [8]

In the real world, a scherzo—from the Italian world for "joke"—is a light-hearted musical composition. That explains why, later in the novel, Tichy tells a character named Professor Trottelreiner, "I've heard, or rather felt, Kitschekov's *Scherzo*, but can't say it had any aesthetic effect on me; I laughed in all the wrong places" [9].

In this particular future (which may or may not be real), composers no longer need to create great music because an audience can be made to think that's what they're hearing through the use of designer drugs. Here's what Trottelreiner says on the subject:

> The people who make something real—they're a vanishing breed. Composers accept their fees, pay their patrons kickbacks, and to the public that comes to the Philharmonium to hear the commissioned work performed they slip a little polysymphonicol contrapuntaline. [10]

Now we're getting really sci-fi, with the suggestion that future "music" might bypass the senses altogether through a purely psychological effect. It's an idea that goes back a surprisingly long way, to a 1945 novella called "The Mule" by Isaac Asimov—which found its way, a few years later, into his novel *Foundation and Empire*. It includes a scene featuring an instrument called a "visi-sonor"—supposedly quite an antique by the time of the story.

Explaining the visi-sonor to the protagonist, Bayta Darell, the character Ebling Mis says: "there are very few really good players; it's not so much that it requires physical co-ordination—a multi-bank piano requires more, for instance—as a certain type of free-wheeling mentality". He goes on:

> All I've made out so far is that its radiations stimulate the optic centre of the brain directly, without ever touching the optic nerve. It's actually the utilization of a sense never met with in ordinary nature. Remarkable, when you come to think of it. What you hear is all right. That's ordinary. Eardrum, cochlea, all that. [11]

They're then treated to a performance by a master of the instrument, Asimov's description of which spans two full pages. The general flavour can be gathered from the following excerpt:

> A little globe of pulsing colour grew in rhythmic spurts and burst in midair into formless gouts that swirled high and came down as curving streamers in interlacing patterns. They coalesced into little spheres, no two alike in colour—and Bayta began discovering things. She noticed that closing her eyes made the colour pattern all the clearer; that each little movement of colour had its own little pattern of sound; that she could not identify the colours; and, lastly, that the globes were not globes but little figures … little shifting flames, that danced

and flickered in their myriads; that dropped out of sight and returned from nowhere, that whipped about one another and coalesced then into a new colour… Beneath it all, the sound of a hundred instruments flowed in liquid streams until she could not tell it from the light. [11]

There's an echo here of a real-world condition called synesthesia (see Fig. 2), which Wikipedia defines as "a perceptual phenomenon in which stimulation of one sensory or cognitive pathway leads to involuntary experiences in a second sensory or cognitive pathway" [12].

The BBC's music website includes a list of "Five Composers with Synesthesia" [13]—and we've already encountered three of them in this book. Franz Liszt was mentioned in the first chapter, as one of the first composers to produce music which attempted to depict the supernatural. Also featured in the first chapter was György Ligeti, whose music was used to such good effect in *2001: A Space Odyssey*—and who popped up again in the previous chapter

Fig. 2 The artist Cassandra Miller has produced a number of "synesthetic" paintings based on musical works, such as this one inspired by Erik Satie's *Gymnopédie* (Cassandra Miller, CC-BY-SA-4.0)

("Science Fiction and Music Culture") with his madcap sci-fi opera *Le Grand Macabre*. Another "synesthetic" composer we met in that chapter was Olivier Messiaen, who was an inspiration for SF author Langdon Jones, and whose work *Chronochromie* hints at his propensity to associate musical sounds with colours.

Also on the BBC's list is the Russian composer Alexander Scriabin (1872–1915)—who, we are told, "was very much influenced by his colour sense, going on to write *Prometheus: The Poem of Fire*, which featured the *clavier à lumières*, a keyboard instrument which emitted light instead of sound" [13].

All these four composers—Liszt, Scriabin, Messiaen and Ligeti—were at the cutting edge of the music of their time, actively pushing its boundaries forward. That can't be said of the BBC's remaining "synesthetic composer", Jean Sibelius (1865–1957), whose music tends to have a nostalgic hankering for an earlier age. In the same year, 1905, that saw Schoenberg's brashly forward-looking *Chamber Symphony*, Sibelius wrote a resolutely old-fashioned piece called "At the Castle Gate". In spite of that, the latter will be familiar to most British space buffs, since it's been used as the theme tune for the BBC's *Sky at Night* since 1957.

In the real world, it's virtually impossible for a synesthetic composer to convey that aspect of their experience to an audience. Attempts to do so with devices like Scriabin's *clavier à lumières* only give a vague hint of what's actually in the composer's head.

It's a different matter in science fiction, because the SF world has the benefit of telepaths—like Gerald Howson, the protagonist of John Brunner's novella "The Whole Man" (1959). He encounters a frustrated young composer named Rudi Allef, and discovers—via telepathy—that this frustration originates in a complex case of synesthesia:

> Rudi Allef's mind was almost as far from the normal as was Howson's own, but in a different direction. Somehow, Rudi's sense-data cross-referred, interchangeably. Howson had trespassed in minds with a limited sort of audio-vision—those of people to whom musical sounds called up associated memories of colours or pictures—but compared to what went on in Rudi's mind that was puerile. [14]

Howson goes on to explain to another telepath, named Clara, why Rudi finds it so difficult to externalize the compositions he experiences in his head:

> Rudi's sensory impressions are so completely interlocked I doubt if he could possibly visualize anything straightforwardly. He hears a note struck on your

piano, and he immediately links it up with—oh, let's say the taste and texture of a slice of bread, the colour of a stormy sky, and the smell of stagnant water, together with a bodily sensation of anxiety and pins-and-needles in the left arm. All those interlock with still other ideas—result, chaos! [14]

Eventually Howson works out a solution, with the assistance of Clara and an aquarium-style tank filled with shifting coloured liquids like a giant lava lamp[3]:

> They had spent the week experimenting, improving and training: now the tank could respond to virtually anything, and they had jury-rigged new controls until it was as versatile and essentially as simple as a theremin. And Clara … was reading Rudi's fantastic mental projections, sifting them out and extracting the essentials, and converting them into visual images, as fast as Rudi himself could think them… Mountains grew in the tank, distorted as if looked at from below, purple-blue and overpowering; mists gathered round their peaks, and an avalanche thundered into a valley surrounded by white sprays of snow, as a distant and melancholy horn theme dissolved in Rudi's mind into a cataclysm of orchestral sounds and a hundred un-musical noises. [14]

In a world where psychic powers really work, the musical possibilities are endless. Another example can be found in Philip K. Dick's novel *The Simulacra* (1964), the very first paragraph of which refers to "the famed Soviet pianist Richard Kongrosian, a psychokineticist who played Brahms and Schumann without manually approaching the keyboard" [15].

Turning that concept around, what about a keyboard instrument that makes a direct psychic connection with the mind of the "listener", without needing to use sound at all? It's an idea that features in another Dick novel, *We Can Build You*, from 1972. The protagonist Louis Rosen and his partner Maury Rock are in the business of making electronic organs, but their competitors have gone a stage further. As Maury says to Louis early in the novel: "Look at the Hammerstein Mood Organ, look at the Waldteufel Euphoria, and tell me why anyone would be content merely to bang out music" [16]. Louis is forced to agree:

> Maury had a point. What had undone us was the extensive brain-mapping of the mid-1960s and the depth-electrode techniques of Penfield and Jacobson and Olds, especially their discoveries about the mid-brain. The hypothalamus is

---

[3] That's what Brunner's description sounds like, although the story was written several years before lava lamps were invented.

where the emotions lie, and in developing and marketing our electronic organ we had not taken the hypothalamus into account. The Rosen factory never got in on the transmission of selective-frequency short range shock, which stimulates very specific cells of the mid-brain, and we certainly failed from the start to see how easy—and important—it would be to turn the circuit switches into a keyboard of 88 black and whites. [16]

At least one of the names in that quote refers to a real person: Wilder Penfield (1891–1976), a neurosurgeon who did indeed carry out important research on the neural stimulation of the brain. He'd already been name-dropped in an earlier Dick novel, *Do Androids Dream of Electric Sheep?* (1968), which features a device called the "Penfield Mood Organ". Unlike the "Hammerstein Mood Organ" in *We Can Build You*, this isn't a pseudo-musical instrument but a handy bedside gadget that acts as an electronic substitute for mood-altering drugs.

The mood organ notwithstanding, *Do Androids Dream of Electric Sheep?*—the novel on which *Blade Runner* was based—does contain its share of musical references, as do many of Dick's other works. We'll look at some of them now—but because they're not, for the most part, concerned with the topic of "future music", they get a new section of their own.

Philip K. Dick, Music Buff

Dick's interest in classical music, like his writing career, can be traced back to a precociously early age. According to his biographer Lawrence Sutin, "Phil dated the start of his writing career at age 12, which he reached on 16 December 1940—a birth date Phil was pleased to share with his idol, Beethoven" [17]. Three years later he took up employment as a sales clerk in a store selling records and radios—"the only job, aside from SF writer, that Phil ever held," according to Sutin [18].

One of the store's regular customers was a local FM radio station, KSMO, and Sutin adds that Dick "claimed, in later years, to have hosted a classical music programme on KSMO, but no one who knew Phil at the time can recall his having been on the radio" [19].

Dick's love of classical music—and his extensive knowledge of it—shows up time and again in his novels. Characters who, in the hands of any other author, wouldn't go near the genre turn out to be passionate aficionados of it. A good example is Rick Deckard, the hardboiled bounty hunter in *Do Androids Dream of Electric Sheep?* (the character played by Harrison Ford in the *Blade Runner* movie). At one point in the novel, Deckard traces an escaped android to an opera house, where he finds the cast rehearsing Mozart's *Magic Flute*. "What a pleasure," we're told, "he loved the *Magic Flute*" [20].

In another novel, *A Maze of Death* from 1970, we learn the following about protagonist Ben Tallchief in the very first chapter:

> In 2105 he had operated the background music system aboard a huge coloniz-ing ship on its way to one of the Deneb worlds. In the tape vault he had found all of the Beethoven symphonies … and he had played the Fifth, his favourite, a thousand times throughout the speaker complex that crept everywhere within the ship, reaching each cubicle and work area. Oddly enough no one had com-plained and he had kept on, finally shifting his loyalty to the Seventh and at last, in a fit of excitement during the final months of the ship's voyage, to the Ninth—from which his loyalty never waned. [21]

In an Author's Foreword to the same novel, Dick says, among other things, that "all material concerning Wotan and the death of the gods is based on Richard Wagner's version of *Der Ring des Nibelungen*, rather than on the origi-nal body of myths" [22].

Another Wagner opera, *Parsifal*, is mentioned several times in Dick's semi-autobiographical novel *VALIS* (1981). The novel's narrator—effectively Dick himself—is fascinated by the supernatural imagery and philosophical dia-logue of the opera, but frustrated by his inability to make rational sense of it:

> *Parsifal* is one of those corkscrew artifacts of culture in which you get the subjec-tive sense that you've learned something from it, something valuable or even priceless; but on closer inspection you suddenly begin to scratch your head and say, "Wait a minute, this makes no sense." [23]

For science geeks, there's a striking moment in the first act of the opera as the monk-like character Gurnemanz leads Parsifal to the castle of the Holy Grail ("with very slow steps", according to the stage directions). Parsifal says "I scarcely tread, yet seem already to have come far", to which Gurnemanz replies: "You see, my son, time here becomes space" [24].

As Dick says in *VALIS*, "Wagner died in 1883, long before Hermann Minkowski postulated four dimensional spacetime … where did Richard Wagner get the notion that time could turn into space?" [25]. It's even more striking when you realize that the German-speaking Minkowski—and of course Einstein too—used the term *Raumzeit*, exactly echoing Wagner's words "*zum Raum wird hier die Zeit*".

As far as science is concerned, the close juxtaposition of the terms "space" and "time" was very much a development of the 20th century, not the 19th. Even more spookily, the phenomenon that Wagner describes, with the passage of time appearing compressed to someone who's moving, exactly matches the Einsteinian concept of time dilation.[4]

VALIS contains references to contemporary music as well as classical, with two of the novel's characters actually working in the music industry. First there's Eric Lampton, described in the novel as a "rock star who is rated with Bowie and Zappa" [26], and fairly clearly based on David Bowie himself. Then there's Brent Mini, who's even more obviously modelled on Brian Eno—one of the pioneers of electronic music we met in the chapter on that subject. According to Sutin, Eno's music was brought to Dick's attention by fellow SF writer K. W. Jeter:

> In late 1977, Jeter introduced Phil to *Discreet Music*, an album of tape loop minimalist music by Brian Eno—the inspiration for the "synchronicity music" of composer Brent Mini in *VALIS*. Phil adored the album, playing it constantly. [27]

Six years after VALIS was published, it had the unusual distinction of being adapted into an opera itself—albeit a rather unconventional one. The work of American composer Tod Machover, it was commissioned by IRCAM—the Paris-based centre for electronic music mentioned in the "Scientific Music" chapter—to mark its tenth anniversary in 1987.

As operas go, *VALIS* is a short one. At less than 80 minutes, it's a third the length of *Parsifal*—and too short to give anything but a brief flavour of Dick's novel. According to Machover, his main goal was "to express psychological states, intellectual ideas and emotional reactions through music" [28]. Apart from the voices of the characters, the score is almost entirely electronic, the only acoustic musicians being a pianist and percussionist.

To reflect the novel's eclectic nature, the opera uses a mix of musical styles. As Machover says, "the great quantity of different types of music in *VALIS* is one of the first things that any listener will notice"—adding that, "the styles of musical expression range from rock to romantic to medieval to futuristic" [28].

Machover himself played the role of Brent Mini—a non-speaking part—at the opera's premiere. Not surprisingly, the music here is reminiscent of Brian

---

[4] Ironically, with a running time of around four hours, *Parsifal* could benefit from being compressed in time itself.

Eno's electronic minimalism—which, in Machover's words, "establishes the feeling that time and forward direction have stopped". On the other hand, he characterizes Eric Lampton's music as "harsh and repetitive rhythms and open, consonant harmonies"—which is basically classical-speak for "rock". At the other extreme, the opera even includes a brief quotation from Wagner's *Parsifal*—both words and, in distorted form, music.

Another Dick novel with strong musical connections is *Flow My Tears, the Policeman Said*, from 1974 (see Fig. 3). The main protagonist, Jason Taverner, is a pop singer who hosts his own TV show, but the novel's eyecatching title refers to another character, Police General Felix Buckman. He's a devotee of the Elizabethan composer John Dowland (1563–1626)—a near-contemporary of Shakespeare who Buckman, in the novel, describes as "the first man to write a piece of abstract music" [29].

The piece in question is one of Dowland's works for solo lute—an instrument that, according to rock musician Sting, is "close enough to the guitar for a modern guitarist to feel relatively familiar with it, but different enough in tuning and fingering to force a brain-teasing restructuring of synapses" [30]. One scene in Dick's novel has Buckman listening to a recording of the Dowland work:

> He stood listening to the *Lachrimae Antiquae* pavane. From this, he said to himself, came, at last, the Beethoven final quartets. And everything else. Except for Wagner. He detested Wagner. Wagner and those like him, such as Berlioz, had set music back three centuries. Until Karlheinz Stockhausen in his *Gesang der Junglinge* had once more brought music up to date. [29]

Fig. 3 *Flow My Tears, the Policeman Said* stands out as one of Philip K. Dick's more unusual titles. It's based on a song written almost four centuries earlier by John Dowland

It seems odd that Buckman is so scathing about Wagner, given Dick's references to the *Ring* and *Parsifal* in other novels. It may be that he liked the stories but not the music—or simply that in this case the character's opinion doesn't reflect the author's. In any case, it's nice to see him say something favourable about Stockhausen, who is one of the main protagonists of the present book.

As for Dowland—it's open to debate whether he really was the first person to write "abstract music". The piece Buckman is talking about is a pavane, which is a kind of dance—and dance music has been around since time immemorial. Then again, the work's title, *Lachrimae*, is Latin for "tears", so perhaps the idea is that it was essentially emotive music that was meant to be sat and listened to rather than danced to.

Dowland went on to produce a vocal version of *Lachrimae* called "Flow My Tears"—which, of course, is where the first half of the novel's title comes from. Interestingly, more than 30 years after the book was written, "Flow My Tears" became a real-world hit when it was featured, along with several other Dowland songs, on the album *Songs from the Labyrinth* (2006) by singer-songwriter Sting, formerly of the band The Police.[5]

Sting, like Buckman in Dick's novel, has a high opinion of Dowland's place in musical history, describing him as "perhaps the first example of an archetype with which we have become familiar, that of the alienated singer songwriter" [30]. While that's impossible to substantiate in a literal sense, it's true that Dowland was one of the first musicians we know who built his reputation on heart-rendingly emotive music. It's a reputation he seems to have been proud of, going so far as to call one of his lute pieces *Semper Dowland, semper Dolens*—Latin for "Forever Dowland, Forever Sad".

As strange as it was for Sting to resurrect Dowland's centuries-old music on *Songs from the Labyrinth*—an album that made the Top 30 on both sides of the Atlantic—it's an idea that was foreshadowed in yet another Philip K. Dick novel. *The Divine Invasion* (1981), set in outer space in the far future, sees fictional recording artist Linda Fox performing Dowland songs to the accompaniment of equally fictional "vibrolutes" [31].

The Divine Invasion was the last novel published before Dick's untimely death in 1982. From the other end of his career comes another piece in which music plays a central role. "The Preserving Machine" was one of his very first short stories, published in the *Magazine of Fantasy and Science Fiction* in June 1953. It concerns an eccentric scientist, Doc Labyrinth, who is worried that

---

[5] "Flow My Tears", the man from The Police said.

civilization is on the point of collapse—and so decides to construct a machine that will ensure the survival of great works of music.

Bizarrely (the story is basically a comedy), the machine accomplishes this by encoding musical scores in the form of living creatures. The exercise is a partial success, in that, after they're released into the environment, the creatures do exhibit the natural survival traits of all living things. What they do, in fact, is undergo rapid evolutionary adaptation to make them better suited to survive. Sadly, the music they were created to preserve is the first casualty, since it becomes distorted beyond recognition.

As well as being an amusing story, "The Preserving Machine" provides an interesting insight into Dick's personal opinions about the music and composers that are referred to. For example, an elegantly rococo string quintet by Mozart is transformed into a bird—"pretty, small and slender, with the flowing plumage of a peacock". But that's just the start:

> Labyrinth went ahead, feeding the music of many composers into the preserving machine, one after another, until the wood behind his house was filled with creeping, bleating things that screamed and crashed in the night. There were many oddities that came out, creations that startled and astonished him. The Brahms insect had many legs sticking in all directions, a vast, platter-shaped centipede. It was low and flat, with a coating of uniform fur. The Brahms insect liked to be by itself, and it went off promptly, taking great pains to avoid the Wagner animal who had come just before. The Wagner animal was large and splashed with deep colours. It seemed to have quite a temper, and Doc Labyrinth was a little afraid of it, as were the Bach bugs, the round ball-like creatures, a whole flock of them, some large, some small, that had been obtained for the *48 Preludes and Fugues*. And there was the Stravinsky bird, made up of curious fragments and bits, and many others besides. [32]

These can all be taken as cartoon-like caricatures of the different styles of the composers: Brahms is stolid and conservative, Wagner is brash and in-your-face, Bach is dispassionate and precise, Stravinsky is polystylistic and unpredictable. The question is—do any of those traits have survival value?

When Labyrinth and the narrator catch up with one of the Bach bugs, they find it has evolved into a ferocious little thing, with nasty poisonous spines. It gets even worse when they come to play back the recording:

> I listened to the music. It was hideous. I have never heard anything like it. It was distorted, diabolical, without sense or meaning, except, perhaps, an alien, disconcerting meaning that should never have been there. I could believe only

with the greatest effort that it had once been a Bach fugue, part of a most orderly and respected work. [32]

At the start of this section, we saw in a quote from a biography of Dick that Beethoven was one of his lifelong idols—and that's worth bearing in mind when we come to the denouement of "The Preserving Machine". In the story, Beethoven's music is represented by a beetle—and it's the only one of the creatures that isn't forced to adapt in order to suit its environment. Instead, the Beethoven beetle adapts its environment to suit itself, by building its own little house. Here is a paragraph from the very end of the story:

At the base of the sycamore tree a huge dun-coloured beetle was building something, putting a bit of mud into place on a strange, awkward structure. I watched the beetle for a time, puzzled and curious, until at last it noticed me and stopped. The beetle turned abruptly and entered its building, snapping the door firmly shut behind it. [32]

So we can take it that Dick considered Beethoven's music to be the most resilient and survivable of all. There's also a strong hint that it has an independent viability of its own, outside the context of human culture. But is that a reasonable assertion, given its origin within that culture? That's the question we're going to look at now.

Musical Relativity

Are musical values relative to a specific human culture, or are they "absolute" in the sense that aliens from another planet would appreciate similar music to ourselves? It's a question that was touched on in the previous chapter, in the context of Langdon Jones's story "The Music Makers" (1965). Set on Mars, its characters discuss whether human music, such as Berg's Violin Concerto, would have had any meaning for the long-dead Martian natives.

As unlikely as they are to have existed in reality, long-dead Martian civilizations are one of the staples of science fiction. An earlier example is A. E. Van Vogt's short story "The Enchanted Village" from 1950. In it, an astronaut named Jenner crash-lands on Mars near the ruins of an unimaginably ancient settlement. Although they're devoid of life, the ruins are pervaded by a strange, unnatural sound:

Faintly, there came to Jenner's ears a thin, high-pitched whistling sound. It rose, fell, faded completely, then came up again clearly and unpleasantly. Even as Jenner ran toward it, the noise grated on his ears, eerie and unnatural… He tried to imagine what an alien culture would want with a mind-shattering noise—although, of course, it would not necessarily have been unpleasant to them. He stopped, and snapped his fingers as a wild but nevertheless plausible notion entered his mind. Could this be music? He toyed with the idea, trying to visualize the village as it had been long ago. Here a music-loving race had possibly gone about their daily tasks to the accompaniment of what was to them beautiful strains of melody. The hideous whistling went on and on, waxing and waning.

Later, when he regains consciousness after passing out, he finds that his perception of the sound has changed:

He woke to the sound of a violin. It was a sad, sweet music that told of the rise and fall of a race long dead. Jenner listened for a while and then, with abrupt excitement, realized the truth. This was a substitute for the whistling—the village had adjusted its music to him! [33]

If that's really what had happened, it would have been a pretty dull story. The fact is, however, the music hasn't changed at all—but Jenner, who now possesses a long tail and snout, has. The village has transformed him into the same biological form as the extinct Martians.

Turning to the glitzy world of mass-market sci-fi, the best known example of "alien" music has to be the cantina scene in the first *Star Wars* movie from 1977 (Fig. 4). The irony here is that this music—like almost everything that composer John Williams has written for the *Star Wars* franchise—is as down-to-earth as it gets, not even making the effort to sound superficially alien. Here's a summary of the scene by media studies expert Seth Mulliken:

Luke Skywalker, after leaving his decimated home, travels with Obi-Wan Kenobi to Mos Eisley, a town referred to as a "wretched hive of scum and villainy". Luke enters a bar, the Mos Eisley Cantina. Inside, we see, for the first time in the film, a large collection of strange and unique aliens and a band playing… The structure of the music itself, the form of the song, does not, in and of itself, connote futurized or otherness. It uses drums, percussion, bass and melodic instruments in a manner completely recognizable to modern western popular music. [34]

In 1989, *Star Wars* and its cantina were deliberately parodied by a rival movie franchise in *Star Trek V: The Final Frontier*. This time—in a scene on the planet Nimbus III—the music sounds genuinely alien, as Mulliken explains:

Unlike the Mos Eisley Cantina, where the music contains familiar ideas to place us as listeners in relation to it … the music here is a series of long synthesizer

Fig. 4 *Star Wars* fans dressed as the musicians from the famous cantina scene (Flickr user Gage Skidmore, CC BY-SA 2.0)

drones, containing no structure we recognize. Western ideas of popular music have no hold in this place, and appear to have never touched it. We can notice that down to the very level of the note, music seems to function differently. In Mos Eisley, the very notion of "upbeat" and "lively", the simplest adjective definitions of music, serve the same function as we understand them. The music is coded according to our western understanding. At Nimbus III, we cannot read even a simple code into the music. [34]

By definition, it's virtually impossible to depict or imagine what alien music might sound like. On the other hand, it's possible to approach the question of "musical relativity" from the opposite angle. What would aliens make of Earth music?

There's an amusing treatment of this topic in Jack Vance's 1965 novel *Space Opera*. The title is a pun—"space opera" being a pejorative term for the lazy kind of science fiction in which outer space is used as a stage-set for tales of romance or adventure that could equally well be set on Earth. Taking the phrase literally, however, Vance's novel centres on a repertory company that takes traditional opera productions to distant planets having no connection with Earth culture other than a basic diplomatic presence.

The humour of the novel lies in the way the company insists on presenting intensely human operas to totally non-human species. For example, on the first planet they visit they put on Beethoven's *Fidelio*—an opera about political prisoners in 18th century Spain—for the cave-dwelling Byzantaurs, who have four arms, two heads, rocky skin and no conception whatsoever of music.

At the opposite extreme, the inhabitants of another planet include the "water-people"—seal-like humanoids with a highly developed musical culture, in which every adult is a musician of one kind or another. The water-people are so sensitive to music that they insist the selected opera—Rossini's *Barber of Seville*—is previewed by a single test subject first.

The guinea-pig is "unfavourably impressed"—as a local Earth diplomat, acting as translator, explains:

> He has noticed a large number of clumsy mistakes… The singers—hmm, a word I don't understand—*bgrassik*. Hmm. Whatever it means it's something the singers do incorrectly when attempting to—another unfamiliar phrase: *thelu gy shlrama* during orchestral implications, which result in faulty *ghark jissu*, whatever that is. Implications might mean overtones. The chord sequences—no, that can't be what he means; chord sequences wouldn't move from north to west.
>
> He listened to the monitor, who now was reading from his notes.
>
> The original antiphony was incomplete. The *thakal skth hg* were too close to the *brga skth gz*, and neither were of standard texture. He found the duet about halfway through interesting because of the unusual but legitimate *grsgk y thgssk trg*. He complains that the musicians sit too statically. He thinks that they should move—hop or jump if they will—in order to blend the music. The work is wild, undisciplined, with too much incorrect—substratum? Perhaps he means legato. In any event he cannot recommend the work to his people until these flaws are overcome. [35]

As amusing as *Space Opera* is, the aliens in it are very much sci-fi aliens rather than realistically imagined ones. If and when extraterrestrials are encountered in the real world, they're likely to be far less human-like in their biology, environment and mental processes.

In this sense of genuinely "alien" aliens, one of the most interesting science-fictional examples is the eponymous entity of Fred Hoyle's 1957 novel *The Black Cloud*. A vast interstellar gas cloud possessing a much higher intelligence than humans, it learns of their existence when a research group on Earth makes contact via radio waves.

At one point in the ensuing discussions, the cloud—which the researchers dub Joe—brings up the subject of music:

> Now I wish to inquire into quite a different matter. I am interested in what you call the arts. Literature I can understand as the art of arranging ideas and emotions in words. The visual arts are clearly related to your perception of the world.

But I do not understand at all the nature of music. My ignorance in this respect is scarcely surprising, since as far as I am aware you have transmitted no music. Will you please repair this deficiency? [36]

A member of the research team named Ann happens to be a skilled pianist, and they duly send Joe a recording of her playing Beethoven's *Hammerklavier* piano sonata, opus 106.

Of the 32 sonatas Beethoven wrote for the piano, the *Hammerklavier* is by far the longest—and the most difficult for a pianist to play. It would be even harder if they tried to play it at Beethoven's indicated tempo of 138 half-notes per minute. That equates to one bar every 0.87 seconds (Fig. 5), which is not only too fast for most pianists, but sounds comically "speeded up" to the listener too.

But would an intelligent cloud of interstellar gas see things the same way? Here's what happens in Hoyle's novel:

So the recording was transmitted. At the end came the message: "Very interesting. Please repeat the first part at a speed increased by 30 per cent." When this had been done, the next message was: "Better. Very good. I intend to think this over. Good-bye."

It just so happens that speeding up Ann's recording by 30 per cent results in the exact tempo specified by Beethoven. As she says, "It makes

Fig. 5 Beethoven's indicated tempo implies that all these notes should be played in just 3.15 seconds. The result sounds ridiculously fast to most people, but the super-intelligent alien of Fred Hoyle's novel *The Black Cloud* prefers it at this speed (public domain image)

me feel a little shivery, although probably it was only some queer coincidence, I suppose" [36].

Aside from the issue of tempo—which was just a highbrow joke on Hoyle's part—there's the bigger question of why the cloud appreciated human-made music in the first place. It's a subject that sparks considerable discussion among the characters:

"It defeats me how music can have any appeal for Joe. After all, music is sound and we've agreed that sound oughtn't to mean anything to him," remarked Parkinson.

"I don't agree there," said McNeil. "Our appreciation of music has really nothing to do with sound, although I know that at first sight it seems otherwise. What we appreciate in the brain are electrical signals that we receive from the ears. Our use of sound is simply a convenient device for generating certain patterns of electrical activity. There is indeed a good deal of evidence that musical rhythms reflect the main electrical rhythms that occur in the brain." [36]

In the real world, a couple of decades after Hoyle's novel, the issue of what aliens might make of Earth music—and other terrestrial sounds—acquired a practical dimension when NASA compiled their "Golden Record" (Fig. 6).

Fig. 6 One of the two "golden records", sent into space on board Voyagers 1 and 2 in 1977 and now on their way out of the Solar System (NASA image)

This was carried on board the Voyager 1 and 2 spacecraft when they were launched in 1977. Their scientific mission was to the outer planets of the Solar System, but their trajectories were such that they would carry on out into interstellar space. NASA's idea was to include an audio message for the benefit of any beings that might come across the spacecraft in the far distant future.

The records take the form of analogue phonographic discs. That may seem old-fashioned in this digital age, but it's actually a much more direct and foolproof way to communicate a sound recording. All the recipients need to know is the correct speed at which to spin the disc—and that's encoded, in a form supposedly understandable to any scientifically savvy civilization, on the cover sleeve.

Music only makes up one section of the "sounds of Earth" to be heard on the records, and it's taken from many different cultures around the world [37]. Interestingly, however, the records feature several of the works mentioned in the Philip K. Dick stories discussed earlier in this chapter:

- An excerpt from Mozart's *Magic Flute*—the opera Rick Deckard watches a rehearsal of in *Do Androids Dream of Electric Sheep?*
- One of Bach's preludes and fugues, which were used to create the Bach bugs in "The Preserving Machine".
- The first movement of Beethoven's Fifth Symphony, which Ben Tallchief uses as background music on a spaceship in *A Maze of Death*.
- Part of the String Quartet in B flat, opus 130—one of "the Beethoven final quartets" mentioned by Felix Buckman in *Flow My Tears, the Policeman Said*.

One piece that doesn't appear on the NASA disc, however, is Beethoven's *Hammerklavier* sonata. That's probably a good thing, because if it was the aliens might think they were playing the record at the wrong speed.

References

1. A. Thayer, *Life of Beethoven* (The Folio Society, London, 2001), p. 213
2. M.M. Scott, *The Master Musicians: Beethoven* (Dent, London, 1974), p. 258
3. L. Biggle, The Tunesmith. If, August 1957, pp. 4–35
4. L. Biggle, Spare the Rod. Galaxy, March 1958, pp. 86–106
5. J. Blish, A Work of Art, in *Galactic Cluster* (Four Square, London, 1963), p. 29–44
6. R. Silverberg, *Three Novels* (Bantam, New York, 1988), pp. 367–369
7. A. Huxley. Sexophones and Synthetic Music. Museum of Imaginary Musical Instruments, http://imaginaryinstruments.org/sexophones-and-synthetic-music/

8. S. Lem, *The Futurological Congress* (Penguin Classics, London, 2017), p. 76

9. S. Lem, *The Futurological Congress* (Penguin Classics, London, 2017), p. 108

10. S. Lem, *The Futurological Congress* (Penguin Classics, London, 2017), p. 112

11. I. Asimov, *Foundation and Empire* (Panther, London, 1962), pp. 106–108

12. Wikipedia article on "Synesthesia". https://en.wikipedia.org/wiki/Synesthesia

13. M. Beek, Five Composers with Synesthesia. Classical-music.com, August 2018. http://www.classical-music.com/article/5-composers-synesthesia

14. J. Brunner, The whole man. Sci. Fantasy **34**, 2–62 (1959)

15. P.K. Dick, *The Simulacra* (Ace, New York, 1964), p. 1

16. P.K. Dick, *We Can Build You* (Fontana, Glasgow, 1977), pp. 7–8

17. L. Sutin, D. Invasions, *A Life of Philip K. Dick* (Harper Collins, London, 1994), p. 34

18. L. Sutin, D. Invasions, *A Life of Philip K. Dick* (Harper Collins, London, 1994), p. 50

19. L. Sutin, D. Invasions, *A Life of Philip K. Dick* (Harper Collins, London, 1994), p. 53

20. P.K. Dick, *Do Androids Dream of Electric Sheep?* (Panther, London, 1972), p. 76

21. P.K. Dick, *A Maze of Death* (Pan, London, 1973), p. 10

22. P.K. Dick, *A Maze of Death* (Pan, London, 1973), p. 3

23. P.K. Dick, *VALIS* (Grafton, London, 1992), pp. 149–150

24. R. Wagner, *Parsifal* (CD booklet, Deutsche Grammophon, 1992)

25. P.K. Dick, *VALIS* (Grafton, London, 1992), p. 45

26. P.K. Dick, *VALIS* (Grafton, London, 1992), p. 157

27. L. Sutin, D. Invasions, *A Life of Philip K. Dick* (Harper Collins, London, 1994), p. 252

28. Tod Machover: VALIS: An Opera (CD booklet, Bridge Records, 1988)

29. P.K. Dick, *Flow My Tears, the Policeman Said* (DAW Books, New York, 1975), pp. 94–95

30. Sting, *Songs from the Labyrinth* (CD liner notes, Deutsche Grammophon, 2006)

31. P.K. Dick, *The Divine Invasion* (Corgi Books, London, 1982), p. 20

32. P.K. Dick, *The Preserving Machine and Other Stories* (Science Fiction Book Club, Newton Abbott, 1972), pp. 7–15

33. A.E. van Vogt, The Enchanted Village, in *Destination Universe* (Panther, St Albans, 1968) pp. 63–77

34. S. Mulliken, *Sounds of the Future* (North Carolina, McFarland, 2010), pp. 88–99

35. J. Vance, *Space Opera* (Coronet Books, London, 1982), p. 85

36. F. Hoyle, *The Black Cloud* (Penguin, Harmondsworth, 1960), pp. 173–175

37. "Music From Earth". https://voyager.jpl.nasa.gov/golden-record/whats-on-the-record/music/

Index

© The Editor(s) (if applicable) and The Author(s),
under exclusive licence to Springer Nature Switzerland AG 2020
A. May, *The Science of Sci-Fi Music*, Science and Fiction,
https://doi.org/10.1007/978-3-030-47833-9

Printed in the United States
By Bookmasters